Dynamics and Monitoring of Railway Vehicle Systems

This book provides cutting edge research into railway vehicle dynamics and the condition monitoring of railway vehicles and track.

Using real world global examples, the book focuses on safety and maintenance from a global perspective. Beginning with an explanation of the practical theory behind dynamics and running safety when passing a curve, the book enables readers to understand the motion and vibration of railway vehicles both in theory and in practice. Presenting examples of real-world phenomena, and a central chapter explaining condition monitoring and Condition-Based Maintenance (CBM), the book demonstrates the mechanisms involved in railway vehicle design and production. It also focuses on safety and riding comfort.

Including case studies, the book will be of interest to railway engineers, and those interested in design, dynamics, condition monitoring and maintenance engineering.

Dynamics and Monitoring of Railway Vehicle Systems

Hitoshi Tsunashima, Akira Matsumoto,
Yohei Michitsuji, Peter Hubbard and
Christopher Ward

CRC Press
Taylor & Francis Group
Boca Raton London New York

CRC Press is an imprint of the
Taylor & Francis Group, an **informa** business

First edition published 2026
by CRC Press
2385 NW Executive Center Drive, Suite 320, Boca Raton FL 33431

and by CRC Press
4 Park Square, Milton Park, Abingdon, Oxon, OX14 4RN

CRC Press is an imprint of Taylor & Francis Group, LLC

ISBN: 9781032422657 (hbk)
ISBN: 9781032422954 (pbk)
ISBN: 9781003362135 (ebk)

DOI: 10.1201/9781003362135

Typeset in Times
by Newgen Publishing UK

Contents

About the Authors

Hitoshi Tsunashima completed his Masters course in Aeronautical Engineering at the University of Osaka Prefecture in 1983. He joined Kobe Steel Ltd. in 1983, where he carried out his research and development of an Automated People Mover. He obtained his Doctoral degree from The University of Tokyo in 1995. He joined the Department of Mechanical Engineering, College of Industrial Technology, Nihon University in 1996 and currently he is a professor and a deputy director of the Centre for Railway Research. His main research expertise includes modeling and simulation of railway vehicles, applications of advanced control for railway vehicles, condition monitoring of railway systems, multi-body dynamics, human factors and ergonomics.

Akira Matsumoto graduated from Yokohama National University with a Bachelor's Degree in Engineering in 1972. He worked in a governmental research institute of the Ministry of Transport from 1972, and then the National Traffic Safety & Environment Laboratory (NTSEL) to 2007. In these laboratories, he was engaged in the research and development of new urban transportation systems, railway safety technologies and their assessment, high curving performance railway bogies and vehicles, wheel/rail interface technologies, especially countermeasures to rail corrugation. In addition, he was involved in the accident investigations of serious railway accidents of the expert Judgments for Police examination on "flange-climb derailment in Tokyo Subway in 2000", "over-turn derailment on Fukuchiyama line of West JR in 2005", and the like. He was a member of Aircraft & Railway Accidents Investigation Commission (ARAIC), and then Japan Transport Safety Board (JTSB) as the chair of the Railway Subcommittee from 2007 to 2016. He obtained his Doctoral Degree from The University of Tokyo in 2014. He was a professor of the College of Industrial Technology, Nihon University from 2017 to 2019. He is the representative director of the Japanese Society of Contact Mechanics of Railways (JSCMR).

Yohei Michitsuji received his Ph.D. from the Department of Industrial Mechanical Engineering, the University of Tokyo in 2004. He joined the Department of Mechanical System Engineering, Faculty of Engineering, Tokyo University of Agriculture and Technology as a lecturer in 2004. He joined the Department of Mechanical Engineering, Faculty of Engineering, Ibaraki University as an Associate Professor in 2009 and is currently a Professor. His main research expertise includes modelling and simulation of railway vehicles, condition monitoring of railway systems, multi-body dynamics, wheel/rail wear and steering bogie design.

Peter Hubbard completed a MEng undergraduate course in Systems Engineering at Loughborough University, followed by a PhD at Loughborough sponsored by BAE Systems researching fault management and reconfigurable control systems in avionics. After a short spell in industry working for the Systems Engineering Innovation Centre with BAE Systems, he joined the School of Electronic, Electrical and Systems Engineering at Loughborough University and researched rail vehicle condition maintenance and mechatronic vehicle guidance. He is now the head of the Control Systems Research group and has research expertise in rail vehicle dynamics, applied mechatronics and applications of MPC in robotics and vehicles.

Christopher Ward completed his Bachelor's Degree in Mechanical Engineering in 2003 and PhD in Automotive Powertrain Control in 2009, both from the University of Liverpool, UK. He moved to Control System Group at Loughborough University, UK, in 2009 as a post-doctoral research associate, becoming a lecturer in 2012 and a senior lecturer in 2019. At Loughborough University his research work focused on rail vehicle condition monitoring and mechatronic guidance concepts. In 2023 he moved to the Rail Accident Investigation Branch as an Inspector of Rail Accidents. In addition to his primary role of investigating accidents he manages the branch's vehicle dynamics work. Chris is a chartered engineer, a member of the Institution of Mechanical Engineers and a Fellow of the Higher Education Academy.

Preface

This book provides a systematic and easy-to-understand explanation of railway system dynamics, focusing on vehicle dynamics and vibration, as well as related condition monitoring. This will be very useful for beginners who are just starting to learn, but also for engineers already involved in the field.

Chapters 1 and 2 provide a detailed explanation of the basics of railway vehicles, allowing the reader to understand how railway vehicles are modelled using elements such as mass, springs and dampers. In addition, the characteristics of wheel-rail systems, including passing through curves, and bogie dynamics are explained in detail. Chapter 3 introduces examples of safety analysis using Multi Body Dynamics (MBD) simulation, such as flange-climb derailment and vehicle overturning.

Chapter 4 introduces the basic concepts of condition monitoring in railways, the methods used for condition monitoring, and application examples. In Chapter 5, we introduce actual examples of the measurement of wheel/rail contact forces, such as derailment coefficients, using in-service trains, and how the results of these measurements are used. We also introduce various methods for estimating track irregularities from measured data by measuring vibrations in the cabin of in-service trains. In addition, we introduce a case study of the development of a method for estimating the adhesion force between wheels and rails in the UK.

This book is characterised by the fact that it does not simply list textbook-style theories, but instead introduces many real-life examples experienced by the authors. This book focuses on examples from Japan that have rarely been introduced before.

It is likely that condition monitoring on the railway will continue to evolve in the future, but in order to carry out effective condition monitoring, it is essential to have a thorough understanding of the dynamics of railway vehicles. This book has been written with this aim in mind, and we hope that it will be widely read by students, researchers and engineers involved in practical work.

Hitoshi Tsunashima
Akira Matsumoto
Yohei Michitsuji
Peter Hubbard
Christopher Ward
March 2025

Acknowledgements

In publishing this book, we would like to express our sincere gratitude to all those involved in the relevant publishing subcommittee of the Japan Society of Mechanical Engineers, especially for allowing us to cite content from Railway Vehicle Dynamics and Modeling, published by the JSME, especially in Chapters 1 to 3. We would also like to thank all those at East Japan Railway Company and Central Japan Railway Company for their cooperation on the condition monitoring technologies using track equipment mentioned in Chapter 4. We are also very grateful to the National Traffic Safety and Environment Laboratory, Tokyo Metro, Nippon Steel Corporation and Nippon Steel Railway Technology for their invaluable assistance in the description of the PQ monitoring bogie described in Chapters 4 and 5. Finally, we would like to thank the CRC press publisher for accepting our proposal and supporting the book writing.

1 Fundamentals of Railway Vehicles and Tracks

1.1 RAILWAY VEHICLES

Railway vehicles are classified as, for example, passenger coaches, freight wagons, diesel locomotives, electric locomotives, diesel multiple units, electric multiple units and so forth. Although the structure of each type of vehicle differs depending on its intended use, there are many similarities in the concept of traction and guidance equipment to ensure safe running. This section describes the basic structure of the most common types of railway vehicles; a vehicle with a pair of bogies with two wheelsets on each.

1.1.1 Coordinate System for Railway Vehicles

When analysing the dynamics of railway vehicles, the wheelsets, bogie frames and a car body that make up the system are often treated as rigid bodies. The Degrees Of Freedom (DOFs) for each rigid body are translational motion along the three axes and rotational motion about the three axes for a total of six DOFs, as shown in Figure 1.1.

1.1.2 Basic Structure of a Railway Vehicle

Figure 1.2 shows the basic structure of a typical railway vehicle with the names and dimensions of each part. The dimensions in the figure are those of a typical conventional railway vehicle and those in parentheses are those of a Shinkansen train. The main dimensions of the bogies are shown in Figure 1.3. In Japan, most bogies have a wheel diameter of 860 mm and the longitudinal distance between the wheelsets in a bogie is about 2 m. Normally, trains are operated with multiple cars connected via couplers, but the vertical and lateral restraints of the car body by the couplers are generally weak, so a single car model is often used for dynamic analysis of a railway vehicle. However, for a Shinkansen train with yaw dampers between the cars, as shown in Figure 1.4(a), or an articulated vehicle with bogies between cars as shown in Figure 1.4(b), a multi-car model should be used. An articulated vehicle or an articulated vehicle with bogies between the car bodies is modelled as a three- or five-car articulated vehicle.

DOI: 10.1201/9781003362135-1

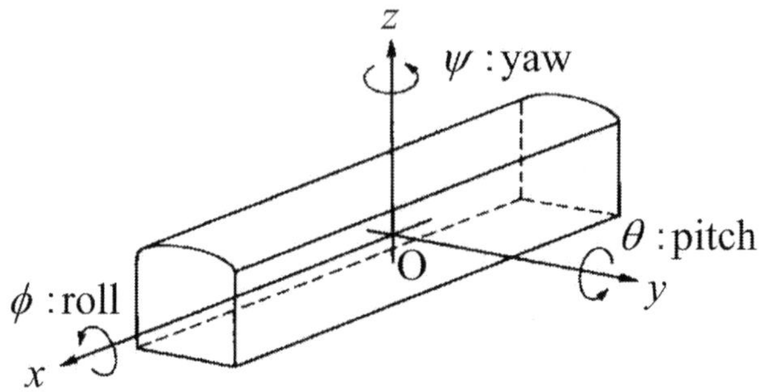

FIGURE 1.1 Coordinate system for a railway vehicle.

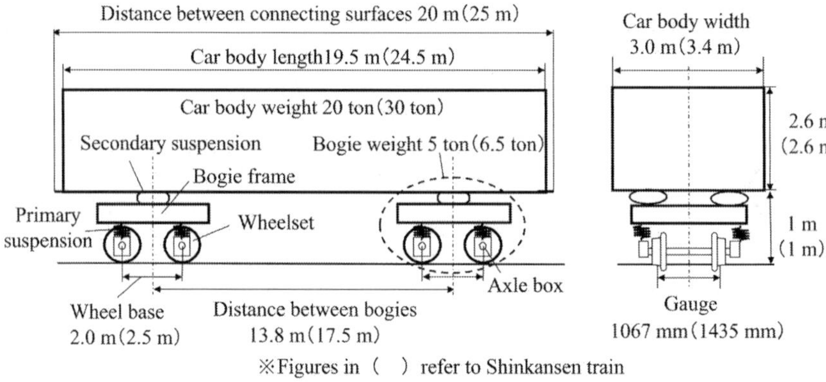

FIGURE 1.2 Basic configuration and main dimensions of a two-axle bogie railway vehicle.

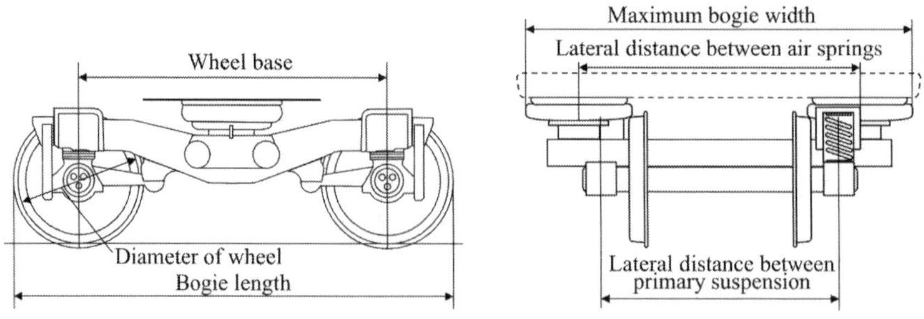

FIGURE 1.3 Definitions of bogie dimensions.

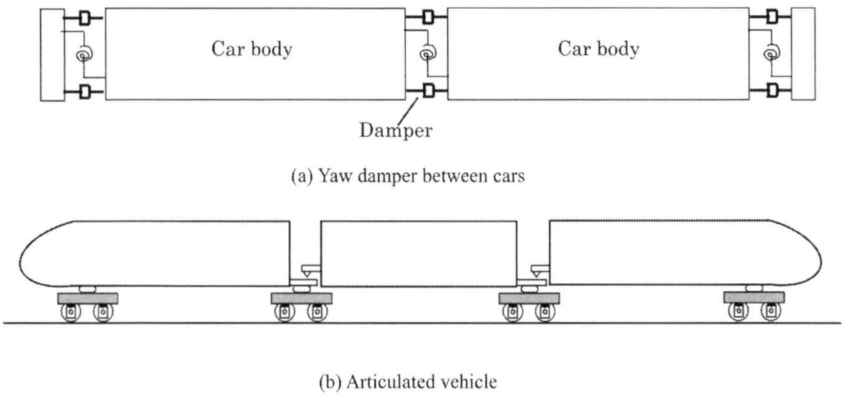

FIGURE 1.4 Example of multiple car body model.

1.1.3 Bogie Function and Structure

1.1.3.1 Role of the Bogie

The railway bogie is crucial for the dynamics of railway vehicles. The bogie of a railway vehicle plays the following four roles:

1. Car body load support
2. Prevention of car body vibration
3. Turning and steering in curves
4. Traction and braking longitudinal force transmission.

The weight of a two-bogie vehicle is supported by four wheelsets. A bolsterless bogie structure is currently in widespread use, as shown in Figure 1.5. The vertical load of the car body is transmitted to the bogie frame, primary spring and wheelset via the secondary suspension, and is supported by the rails directly under each wheel. The same elements also contribute to decreased vibration in the car body. For example, wheelset vibrations generated by the rails with track irregularities are reduced from being transmitted to the car body by the damping action of the primary spring and secondary air spring. To damp vibrations, damping elements such as lateral dampers and vertical dampers are often installed in parallel with the springs. When designing a bogie from the perspective of vibration prevention, the bogie and car body are treated as a mass-spring-damper system and their vibration isolation performance is evaluated. In this case, the part below the primary spring, such as the wheel axle, is called the unsprung mass, and the part between the primary spring and secondary spring, such as the bogie frame, is called the intersprung mass. The car body suspension and axlebox suspension are crucial for turning and steering in curves and transmission of longitudinal forces. Traction and braking are also important dynamics of a railway vehicle. In the modelling of the

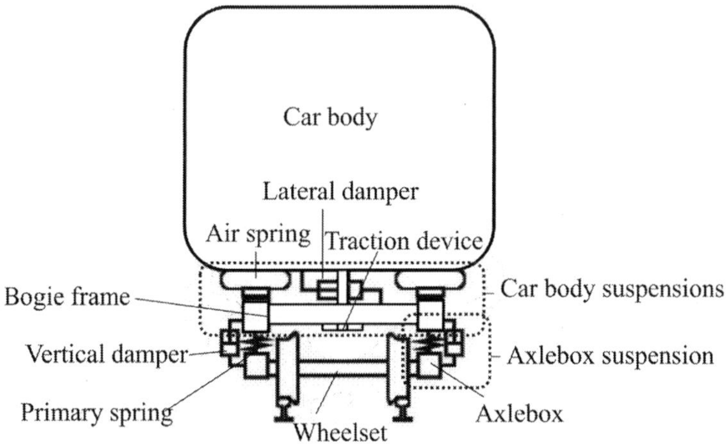

FIGURE 1.5 Basic structure of a railway vehicle (example of bolsterless bogies).

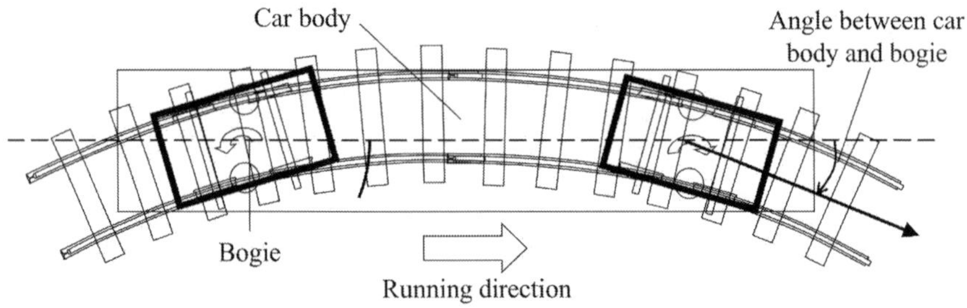

FIGURE 1.6 Bogie angle when passing through a curve.

longitudinal force transmission, the effect of the vertical and lateral dynamics of a railway vehicle is sometimes omitted.

1.1.3.1.1 Bogie with Bolsters

For a bogie supporting a long car body to turn in a curve, it is necessary to allow a large angle between the bogie and the car body. This relative yaw angle between the bogie and the car body is shown in Figure 1.6. For this reason, a bolster is placed between the bogie frame and the car body, and bogies with a structure that turns between the bogie frame and bolster or between the bolster and car body around the centre pivot have been widely used. The former is called a direct-mount bogie because the car body is placed directly above the secondary spring, while the latter is called an indirect-mount bogie because the secondary spring is placed above the bolster and the car body is placed on top of it. The actual structure of a direct-mount bogie is shown in Figure 1.7. A schematic diagram is shown in Figure 1.8(a), and a schematic diagram of the structure of an indirect-mount bogie

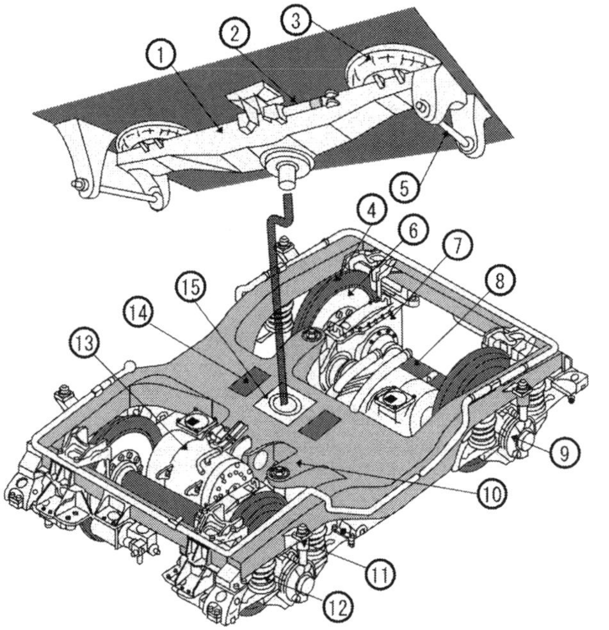

1 Bolster
2 Lateral damper
3 Air spring
4 Wheel
5 Bolster anchor
6 Brake disk
7 Gear box
8 Axle
9 Axlebox
10 Bogie frame
11 Vertical damper
12 Primary spring
13 Traction motor
14 Side bearer
15 Centre plate

FIGURE 1.7 Bogie with bolsters (direct-mount type).

is shown in Figure 1.8(b). In both systems, the bogie can turn at a large angle by turning around the centre plate when passing through a curve, but the secondary spring is not deformed back and forth. To prevent hunting at high speeds, sliding parts such as side bearers are installed to provide turning resistance between the car body and bogie using friction.

Wheelset hunting motion refers to the unstable self-induced vibration inherent to railway vehicles, and it is described in detail in Chapter 2. In this type of bogie, the longitudinal force between the bogie and the car body during traction and braking is transmitted by a bolster anchor installed in the secondary spring section. The bolster anchors are made of steel with rubber bushings inserted at both ends. They are connected longitudinally between the car body and bolster in a direct-mount bogie, and between the bolster and bogie frame in an indirect-mount bogie. When the side bearings do not slide due to static friction or when the yaw angle is small, such as during straight running, the rubber bushings of the bolster anchors function as restoring springs between the car body and the bogie. The traction rod is installed as close as possible to the axle centre height to avoid inducing bending vibration of the car body.

Air springs (pneumatic suspension) with a height adjustment device are widely used as secondary springs. Air springs not only gently support the car body but also provide a damping force against vertical vibration by means of an orifice between the air spring and the surge reservoir, as shown in Figure 1.9. A diaphragm-type

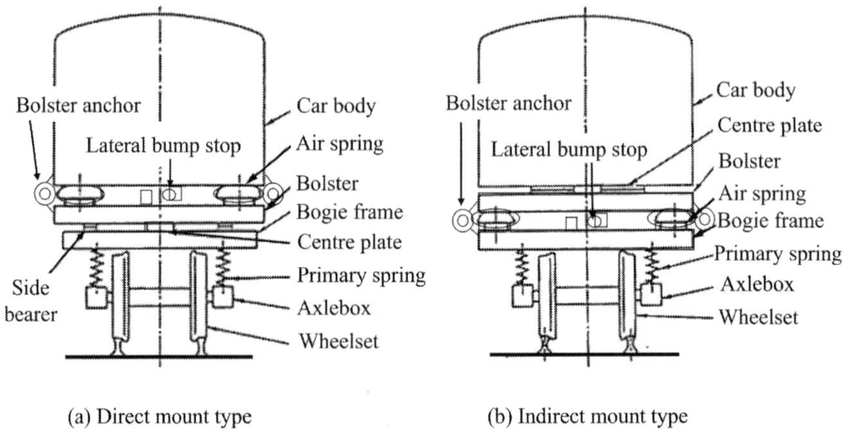

FIGURE 1.8 Schematic diagram of the body support system of a bogie with bolsters.

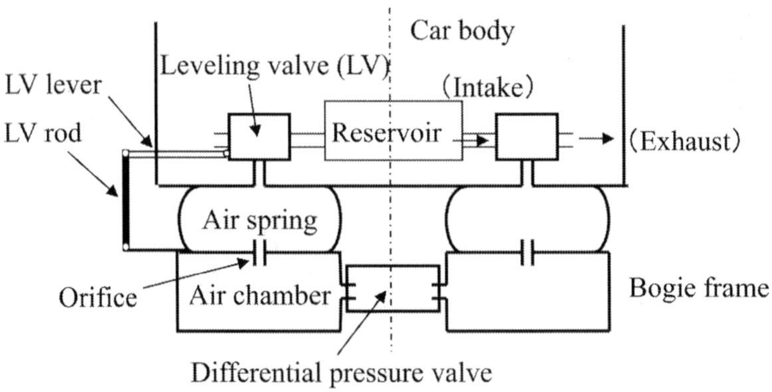

FIGURE 1.9 Configuration of air spring system.

air spring functions as a spring not only in the vertical direction but also in the longitudinal and lateral directions. However, because air springs generate little damping force for vibrations other than those in the vertical direction, oil dampers are installed in parallel. If the rigidity of the vehicle body against roll is insufficient, a torsion spring type anti-roll bar may be added.

To prevent excessive displacement of the secondary spring section, a lateral bump stop is installed between the car body and the bogie. When the relative lateral displacement of the two exceeds a pre-determined distance, the car body-end strikes the stopper rubber attached to the bogie side. In the vertical direction, and the cushioning rubber in the air spring functions as a bump stop. In bogies with bolsters, a steel liftstop is attached to the air spring to prevent the car body from rising past a predetermined limit due to excessive extension of the air spring.

In addition, bogies with bolsters also include swing hanger bogies that use coil springs or bellows air springs as sleeper springs to cushion vertical movement and swing sleepers to provide restoring force for lateral movement of the car body.

1.1.3.1.2 Bolsterless Bogie

In recent years, bolsterless bogies, which are lighter in weight due to the elimination of the bolster, have entered the mainstream for passenger car bogies. Figure 1.10 shows the structure of a bolsterless bogie. When passing through a curve, the bogie rotates around the centre pin, and the air spring deflects largely in the lateral and longitudinal directions when the bogie angle is generated. The development of air springs capable of such a large displacement made the bolsterless bogie a practical reality.

Because the bolsterless bogie eliminates friction elements such as the bolster anchors and side supports, a yaw damper that generates a damping force to suppress hunting behaviour is especially required in high-speed vehicles. The yaw damper is an oil damper that generates a large damping force with a small stroke, and it is usually installed on both sides of the bogie to couple the car body and bogie frame in the longitudinal direction. When the bogie generates a change in yaw angle at a low rate, such as when the vehicle enters a circular curve from a straight track through a transition curve, the yaw damper generates little damping force because the force is proportional to yaw rate. Like the bolster anchor, the yaw damper is

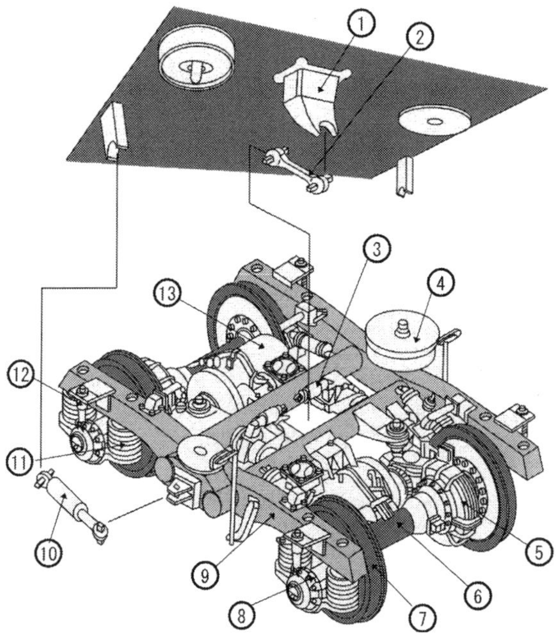

FIGURE 1.10 Example of bolsterless bogie.

installed as close as possible to the axle centre height of the wheelset to avoid inducing bending vibration of the vehicle body.

The longitudinal force is transmitted from the bogie frame to the car body by a traction device located at the centre of the bogie so as not to interfere with the turning motion of the bogie. There are various types of traction devices, including the single-link type, Z-link type, gantry-type and plate spring type. Many bolsterless bogies use a traction device to stop abnormal lift due to the large displacement of the air springs. For example, a bogie with a single-link type traction device has a structure in which the rod of the link hits the bogie frame when the air spring extends, and the car body rises above an allowable limit.

1.1.3.2 Axle-Supporting Device

The axle-supporting system connects the wheelset axlebox, which contains bearings, to the bogie frame. This in turn supports the vertical load and holds the two pairs of axleboxes parallel and at the appropriate position in relation to the bogie frame. They also provide adequate elasticity vertically, but also longitudinally and laterally to prevent wheelset hunting motion. If the axlebox longitudinal and lateral stiffness is too large, the turning performance will deteriorate, while too small a rigidity will cause an unstable hunting motion. In many cases, a model with springs and dampers in the longitudinal, lateral and vertical directions between the axlebox position and the bogie frame is used in vehicle motion analysis. Specific structural types include the pedestal guides, cylindrical guides (Schlieren), trailing arms, leaf springs, beam links, mono-links, and laminated rubbers. Recently, however, the upper and lower loads are supported by a coil spring and the longitudinal and lateral supports are supported by rubber bushings. In addition, link-type bogies and axle beam-type bogies, in which the vertical load is supported by coil springs and rubber bushings are used for longitudinal and lateral support to eliminate the free space between the bogie frame and axlebox, are increasingly used. Some bogies are also fitted with an oil damper in parallel with the primary spring to dampen vertical body vibration. Examples of various axle supporting devices are shown in Figure 1.11. For more detailed information regarding the railway vehicle structure, please refer to other sources [1, 2].

1.1.3.3 Wheels and Wheelsets

The characteristic feature of a railway vehicle is the wheelset, which is assembled by press fitting steel wheels onto an axle. The part of the wheel that comes into contact with the rail surface is called the wheel tread, which has an inclination known as conicity. Figure 1.12 shows the key part of a wheel profile. When the wheelset is displaced laterally from the track centre, the rolling radius at the wheel-rail contact point changes for the left and right wheels due to the profile gradient. When the left and right wheels rotate at the same speed, the so-called self-steering ability of a wheelset occurs. On curves, the vehicle travels along the track using this self-steering ability. To prevent derailment, flanges are provided on the inner surface of the wheels to assist the turning of the bogie on sharp curves where the

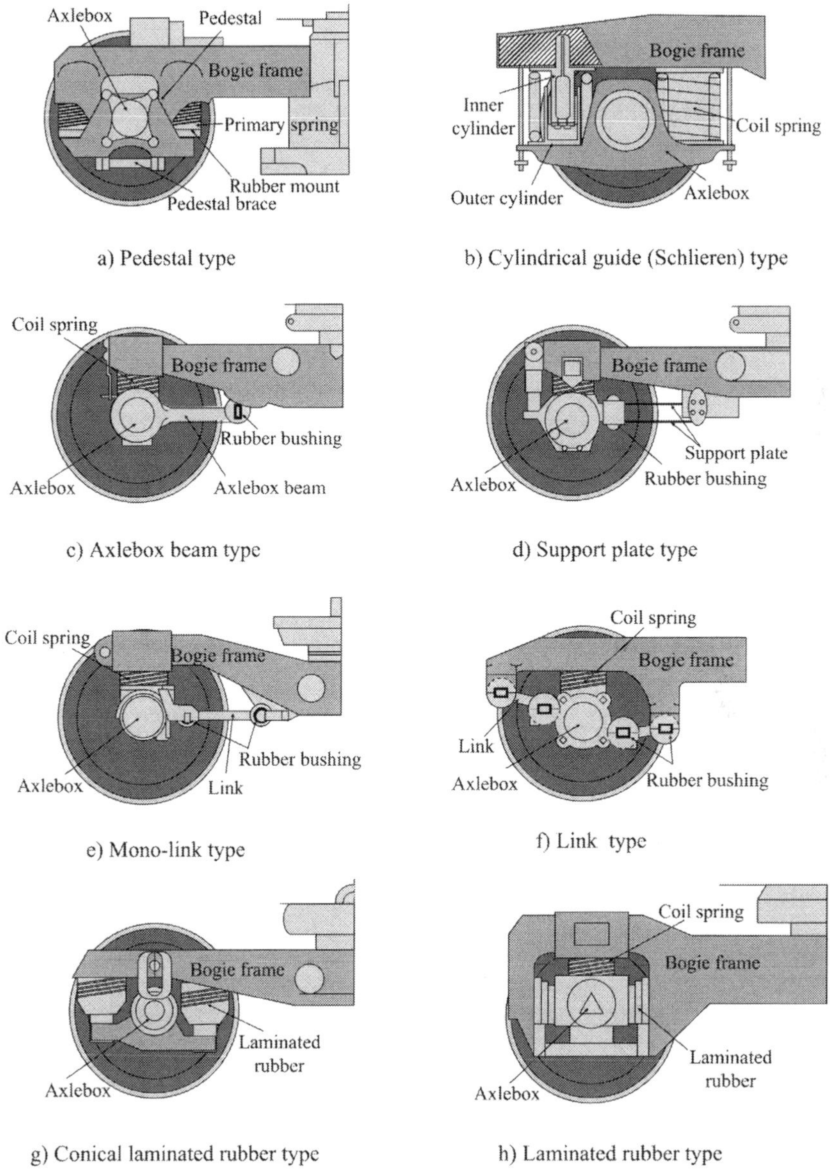

FIGURE 1.11 Examples of various axle support systems.

difference in wheel radii between the left and right wheels is not enough to turn the bogie.

Wheels widely used in Japan have a rim width of 125 mm, and the tread centre is 65 mm from the wheel. The back-to-back gauge is 990 mm for narrow-gauge

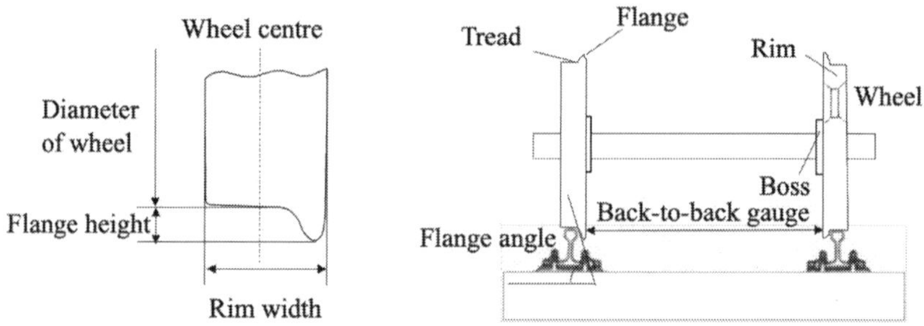

FIGURE 1.12 Definitions of each part of the wheelset.

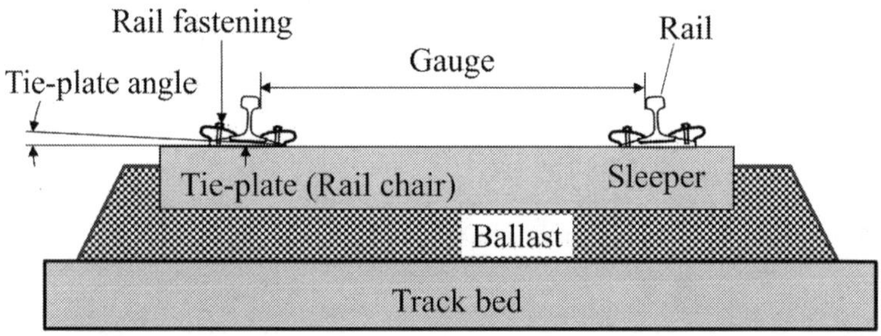

FIGURE 1.13 General track structure.

conventional trains and 1360 mm for Shinkansen trains, with a flange height of 27–30 mm and a flange angle of 65–70°. The tread of the wheel used to have a constant gradient of 1/20 for conventional lines and 1/40 for Shinkansen lines, but currently, wheels with an arc profile are used to achieve both running stability and turning performance. The wheel tread profile and wheel-rail contact are explained in Section 2.1.

1.2 TRACKS

Railway vehicles are guided by the track, which supports and guides the train, ensuring smooth operation. This section provides essential information about railroad tracks, focusing on their geometry, which is crucial for analysing the dynamics of rolling stock.

1.2.1 Track Structure

As shown in Figure 1.13, the track is a structure consisting of rails, sleepers and track beds, which distribute and transmit the train load to the track bed and the structures below. The track bed ballast plays an important role in softening the dynamic train loads transmitted from the sleepers and distributing them widely to the track bed. In Japan, the two rails on each side are fastened to the sleeper with a 1/40 inclination, known as the tie-plate angle. Hard rubber pads called track pads are inserted between the rails and the sleeper and between the tie plate and the sleeper. Track with sleepers laid on ballast track beds is the most common type of track, but in recent years, slab track and track directly connected to elastic sleepers, which are designed to reduce maintenance labour, have been increasingly laid. The distance between the left and right rails is the most basic design criterion for railways and is called the gauge. Gauge is defined as the shortest distance between the inner surfaces of the left and right rails within a specified distance below the rails (often 14–16 mm). As the typical gauges used in Japanese railways, JR conventional lines use narrow gauge (1067 mm) and Shinkansen lines use standard gauge (1435 mm).

1.2.2 Parameters of Track

1.2.2.1 Curve Radius

When laying a curved track, a circular curve is generally used, and the degree of curvature is expressed by the radius *R* (in metres) of the curve at the centre of the track, such as R600. The smaller the curve radius, the more difficult it is to run at high speeds and the greater the risk of derailment. According to Japanese national standards, the minimum curve radius is 160 m for conventional trains, and curves with radius smaller than this are permitted, taking into account the running performance of the train and safety equipment such as check rails. Shinkansen trains require larger curve radii to facilitate high-speed running, with the minimum curve radius set at 2500 m for the Tokaido Shinkansen and 4000 m for the Sanyo Shinkansen and subsequent lines.

1.2.2.2 Cant

When a train passes through a curve, centrifugal forces act on the vehicle, and the higher the speed, the more the vehicle tends to overturn. Therefore, to counteract this centrifugal force with the gravitational component, the outer rails (hereafter referred to as the 'high rail') are raised in relation to the inner rails (hereafter referred to as the 'low rail') on a curve on the main line. This height difference between the high and low rails is called a cant, as shown in Figure 1.14. The maximum amount of cant is limited by the need to ensure that vehicles are sufficiently safe from overturning when stopped on curves and to prevent passengers from experiencing discomfort. The maximum cant is 105 mm on narrow-gauge conventional lines with a gauge of 1067 mm, 140 mm on standard-gauge conventional lines with a gauge of 1435 mm, 160 mm on some private railways and 200 mm on Shinkansen lines in Japan.

On conventional lines, the low rail is kept at the same height as the straight section, while the high rail is adjusted so that the value reaches the specified cant at a circular curve, as shown in Figure 1.15(a). A transition curve inserted

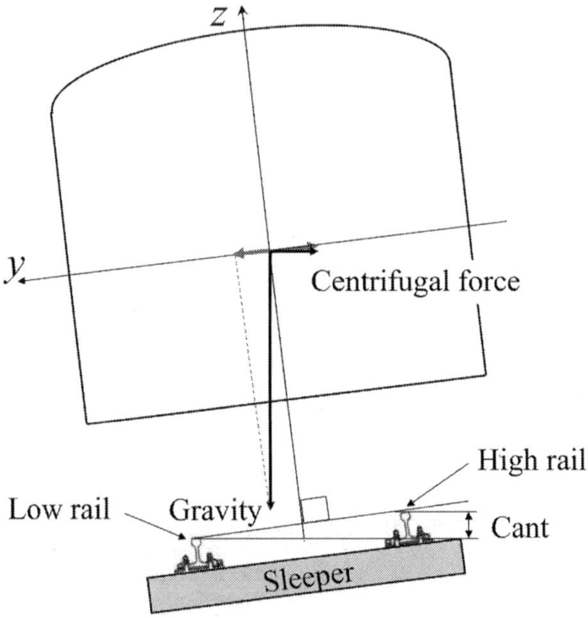

FIGURE 1.14 Cant of a curve.

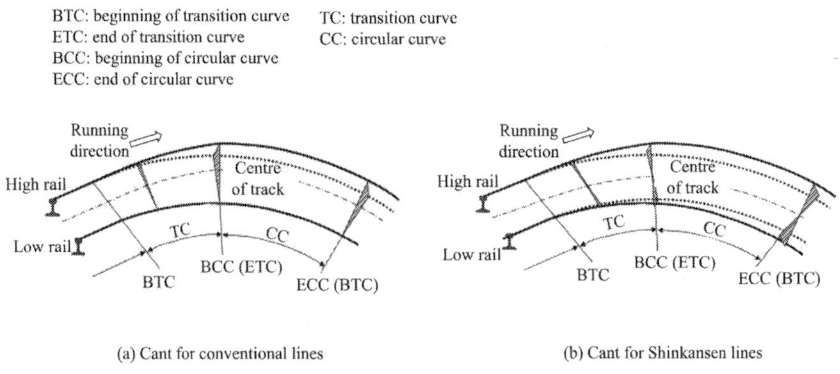

FIGURE 1.15 How the cant is set.

between a straight line and a curve has a gradient as a vertical alignment, and the track centre gradually rises on the entrance transition curve and gradually falls on the exit transition curve. In contrast, for the Shinkansen lines, the low rail is lowered, and the high rail is raised without changing the height of the track centre, as shown in Figure 1.15(b), so there is no gradient in the track centre in the transition curve.

1.2.2.3 Transition Curve

When a track is constructed with a combination of straight sections and constant radius curves, vehicles passing through these transitions would experience a sudden increase in centrifugal force, leading to a significant deterioration in both operational safety and ride comfort. To prevent this, a transition curve is usually inserted between a straight line and a curve. Japanese railways employ cubic parabolic transition curves for conventional lines and half-wavelength sine wave transition curves for Shinkansen lines. The half-wavelength sine shape is advantageous in decreasing vehicle roll motion at high speeds, because the cant gradient varies continuously at both ends of the transition curve. Conversely, for the same transition curve length, the steepest cant gradient in the centre of the curve is larger than that of a linearly changed transition curve. Therefore, the half-wavelength sine wave transition curve is mainly used on Shinkansen lines.

1.2.2.4 Slack

In curves and turnouts with small radii, the gauge is widened to facilitate smooth running of wheelsets. This is called 'slack', and it is achieved along the entire length of a curve by widening the inner gauge with respect to the outer gauge, as shown in Figure 1.16. Currently, the primary purpose of introducing slack is to enable the wheelset to utilise its self-steering capability by creating a differential in the wheel radius between the high and low rail wheels. In practical applications, the amount of slack in a narrow-gauge conventional line is 15 mm on an R200 m curve, 10 mm on an R300 m curve, and 5 mm on an R400 m curve.

1.2.3 Track Irregularities

The shape of the rail, which is the surface on which the vehicles run, gradually changes due to the effects of repeated train traffic and natural phenomena. This is called track irregularity. The track bed ballast is gradually plastically deformed by the repeated dynamic train loads transmitted from the sleepers. This results in

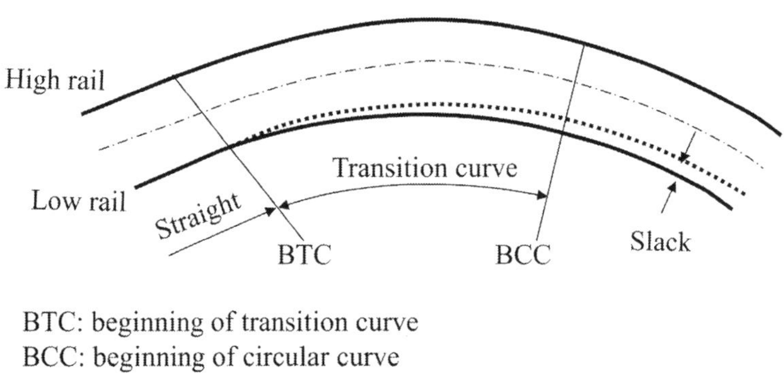

FIGURE 1.16 How slack is added to a curve.

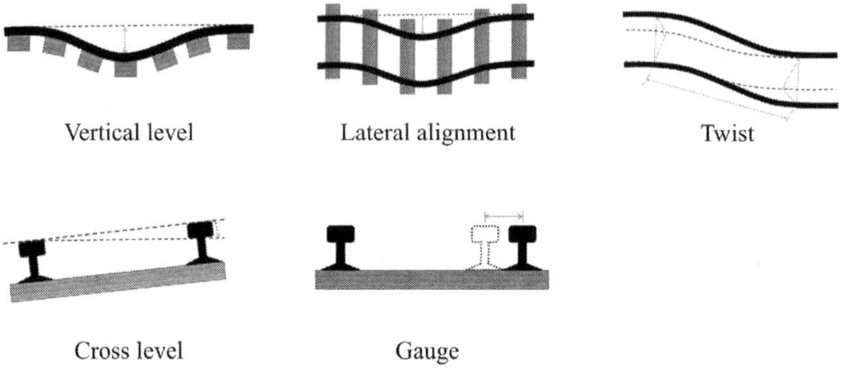

FIGURE 1.17 Types of track irregularities.

residual displacement of the sleepers and rails supported by the track bed. Other types of track irregularity are caused by deformations of the track and structures due to natural phenomena such as rainfall and earthquakes. Track irregularities hinder the smooth running of trains, degrade ride comfort due to vehicle sway, increase lateral force and increase the derailment risk due to a decrease in the wheel load. In addition, the track materials deteriorate due to the significant wheel load and lateral force, so it is important to inspect the track regularly to identify any issues and conduct track maintenance work. Types of track irregularities, including vertical alignment, lateral alignment, cross level, twist, gauge are shown in Figure 1.17. Cant is also monitored in the UK.

Vertical irregularity is the vertical irregularity of the rail head surface. In a track on a track bed, it is mainly caused by the settlement of the track bed ballast. Generally, a 10-meter string is stretched across the top of the rail head and the vertical distance between the rail and the displacement at the centre of the string is measured. Vertical irregularities are associated with the vertical movement of the car body and increase as the variation in wheel load increases.

Lateral alignment refers to the lateral displacement of the rail. Similar to vertical irregularity, it is commonly expressed as 10-meter chord versine values. Lateral alignment contributes to increased lateral forces and train movement in the lateral direction.

Cross level irregularity refers to the height difference between the left and right rails over a given gauge length (for example, 1067 mm for narrow-gauge lines).

Gauge irregularity includes changes such as widening or narrowing. Widening of the gauge may lead to derailment within the gauge, while narrowing of the gauge may lead to derailment due to increased lateral forces. On sharp curves and turnouts, slack is set to ensure smooth running of the vehicle, and the gauge is managed in consideration of this slack.

Twist refers to the change in level over a fixed distance and describes the twist of the track with respect to the plane of the track. Rail vehicles generally have multiple wheelsets fixed to the bogie, and a large twist in the plane can easily decrease the wheel load, so this is an important track irregularity management item from the viewpoint of running safety.

Other irregularities include a combination of alignment and cross level irregularities, which are considered in measures aimed at preventing multiple-factor derailments of freight wagons. For each of these items, track maintenance standards and acceptable standards are set by the railway operator. Track irregularities are inspected by static manual inspection as well as dynamic inspection by track inspection vehicles, which are controlled to the nearest millimetre.

REFERENCES

1. Japanese Society of Mechanical Engineering (JSME), *Railway Vehicle Dynamics and Modelling*, Maruzen (in Japanese), 2017.
2. A. Orlova, Y. Boronenko, The anatomy of railway vehicle running *gear, Handbook of Railway Vehicle Dynamics*, CRC Press, Taylor & Francis, 39–84, 2006.

2 Modelling the Fundamental Dynamics of Railway Vehicles

2.1 CONTACT FORCE BETWEEN THE WHEEL AND RAIL

The dynamics of a rail vehicle, especially the lateral and yaw motion, is strongly influenced by the creep force acting on the contact patch between the wheels and the rail. This creep force is a tangential force resulting from the slip (creep) between the rolling wheel and the rail, and its magnitude is dependent on the creep ratio.

Figure 2.1 illustrates the creep forces acting in the contact patch. In addition to the normal force, the contact patch is subject to a combined longitudinal and lateral creep force caused by the lateral, longitudinal creep and a spin moment around the normal axis of the contact patch [1, 2]. The creep force model, based on the book *Three-Dimensional Elastic Bodies in Rolling Contact* [3] and its related manual [4], is widely used in vehicle dynamics analysis.

Figure 2.2 illustrates the relationship between creep ratio and creep force. The contact patch is elastically deformed [5] in the longitudinal direction and the axis of spin moves forward from the centre of the contact patch, resulting in mutual influence between lateral creep and spin creep. In the range where the creep ratio is small, the creep force increases almost in proportion to the creep ratio. This range is particularly called the linear creep range, and linear creep forces are used when conducting vehicle running stability analysis.

The linear creep force is calculated by the following equation:

$$\left.\begin{array}{ll}\text{Longitudinal creep force:} & f_1 = -\kappa_{11} \nu_1 \\ \text{Lateral creep force:} & f_2 = -\kappa_{22} \nu_2 - \kappa_{23} \omega_3 \\ \text{Spin moment:} & M_3 = \kappa_{23} \nu_2 - \kappa_{33} \omega_3 \end{array}\right\}, \tag{2.1}$$

where $\kappa_{11}, \kappa_{22}, \kappa_{23}$, and κ_{33} are linear creep coefficients of Kalker and $\nu_1, \nu_2, and\ \omega_3$ are the longitudinal, lateral and spin creep ratios, respectively.

The spin moment M_3 is sufficiently small compared to the moment due to the creep force acting on the wheelset, and it is often ignored when analysing vehicle

DOI: 10.1201/9781003362135-2

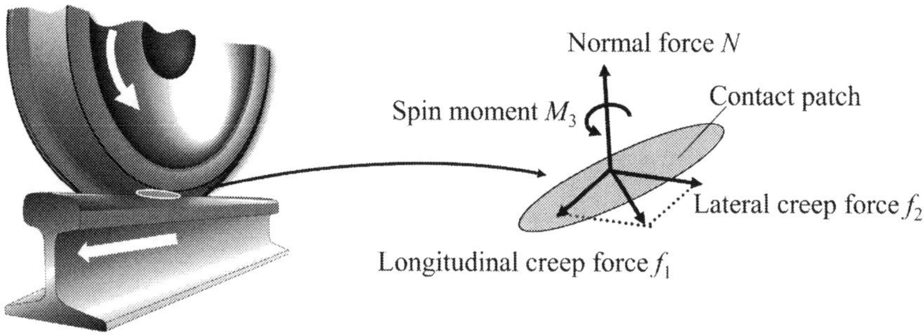

FIGURE 2.1 Creep forces acting on the contact patch between the wheel and rail.

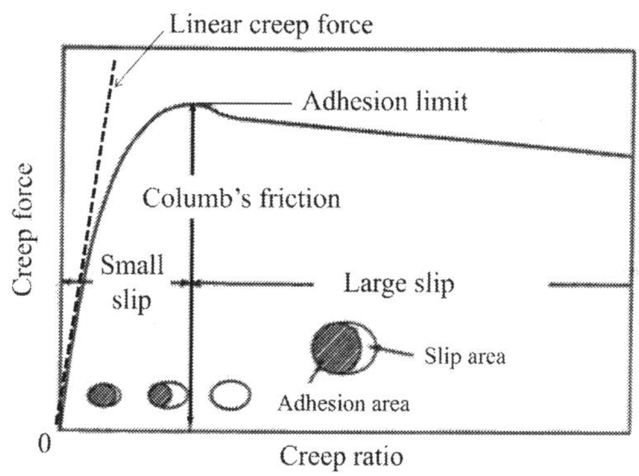

FIGURE 2.2 Creep ratio and creep force between the wheel and rail.

dynamics [6]. The spin term of the lateral creep force $-\kappa_{23}\omega_3$ is also considered in the analysis of a flange climb derailment where the flange contacts the rail and the spin becomes large, but it is often ignored in other normal straight and curved section analyses because it is sufficiently small compared to the term due to lateral creep $-\kappa_{22}\nu_2$. In other words, the following Equation (2.2) is used in most cases for vehicle motion analysis:

$$\left.\begin{array}{ll}\text{Longitudinal creep force:} & f_1 = -\kappa_{11}\nu_1 \\ \text{Lateral creep force:} & f_2 = -\kappa_{22}\nu_2\end{array}\right\}, \tag{2.2}$$

where κ_{11}, κ_{22} are linear creep coefficients and ν_1, ν_2 are longitudinal and lateral creep ratios.

2.2 DYNAMICS OF A WHEELSET

This section deals with the fundamental dynamics of a wheelset based on linearised equations of motion. Even within the framework of a simplified linear model, it is possible to capture the key characteristics of the wheelset, such as its hunting behaviour and self-steering properties.

2.2.1 Equations of Motion for a Wheelset

To analyse the motion of a wheelset, a wheel with a conical tread, as depicted in Figure 2.3, is employed for simplicity, enabling the derivation of the linearised equations of motion. In this figure, both lateral and yaw displacements are also defined. The wheel-rail contact angle is assumed to be small, and it is further considered that the spin moment is negligible compared to the moment due to the longitudinal creep force. The tread gradient (conicity) of the wheel is denoted as γ, and the wheel radius linearly changes with lateral displacement. The distance from the axle centre to the contact point of the left and right wheels is denoted as b_0 if the respective changes are assumed to be small. Under these assumptions, the creep ratios for the left and right wheels are as follows:

Left wheel:

$$\left.\begin{aligned} v_{1L} &= -\frac{\gamma}{r_0} y_W - \frac{b_0}{v}\dot{\psi}_W \\ v_{2L} &= -\psi_W + \frac{\dot{y}_W}{v} \end{aligned}\right\} . \tag{2.3}$$

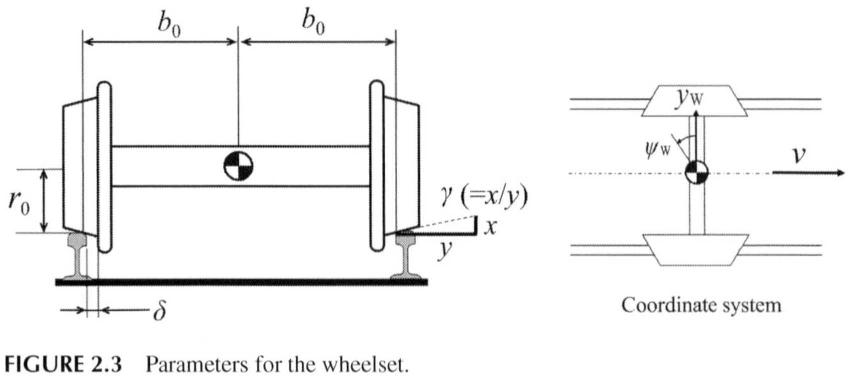

FIGURE 2.3 Parameters for the wheelset.

Right wheel:

$$\left.\begin{aligned} \nu_{1R} &= \frac{\gamma}{r_0} y_W + \frac{b_0}{v} \dot{\psi}_W \\ \nu_{2R} &= -\psi_W + \frac{\dot{y}_W}{v} \end{aligned}\right\}. \tag{2.4}$$

When the creep forces of the left and right wheels are calculated according to Equation (2.2), the yaw moment M_ψ around the wheel centre of gravity due to the lateral creep force Y and longitudinal creep force acting on the wheelset can be expressed by the following equations:

$$\text{Lateral creep force: } Y = f_{2L} + f_{2R} = 2\kappa_{22}\left(\psi_w - \frac{\dot{y}_W}{v}\right),$$

$$\text{Yaw moments: } M_\psi = -b_0 f_{1L} + b_0 f_{1R} = -2\kappa_{11}\left(\frac{\gamma b_0}{r_0} y_W + \frac{b_0^2}{v}\dot{\psi}_W\right).$$

The equations of motion for the wheelset can be expressed as follows:

$$\left.\begin{aligned} m_W \ddot{y}_w + \frac{2\kappa_{22}}{v}\dot{y}_W - 2\kappa_{22}\psi_w = 0 \\ m_w i_w^2 \ddot{\psi}_w + \frac{2\kappa_{11} b_0^2}{v}\dot{\psi}_w + \frac{2\kappa_{11} b_0 \gamma}{r_0} y_w = 0 \end{aligned}\right\}, \tag{2.5}$$

where m_W is the mass of the wheelset and i_w is the yaw inertia radius of the wheelset.

Figure 2.4 shows a block diagram of Equation (2.5). The diagram can be divided into lateral and yaw dynamics because it is a 2-DOF system. Creep force acts as damping force in the lateral and yaw dynamics as shown in term ① and term ④, respectively, in the figure. There is no damper between the wheels and rails, and the creep force between the wheels and rails dissipates energy. The yaw angle is converted into a force in the lateral direction with term ③. In addition, ② in the figure shows that there is a term in which the lateral displacement is converted into a yaw moment, which is a restoring force that tries to return the wheelset to the neutral position. The block diagram clearly shows how, once lateral displacement occurs, it excites yawing, which is converted into a force that generates lateral motion again through the generation of yaw angle. Incidentally, if the left and right wheels are independently rotating wheelsets, the longitudinal creep coefficient κ_{11} can be regarded as zero, and terms ② and ④ in the figure will no longer exist.

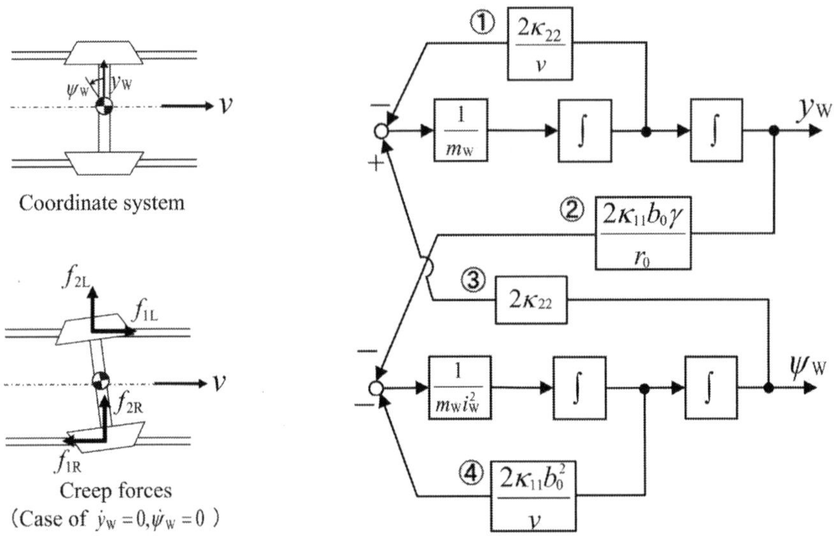

FIGURE 2.4 Dynamics of a wheelset represented by a block diagram.

When the coupling term ② is eliminated, the dynamics coupling is broken, and both the self-steering function and the hunting behaviour are eliminated.

2.2.2 Hunting Motion of a Wheelset [7–10]

Although the stability of a wheelset can be assessed through eigenvalue analysis based on the equations of motion outlined in the previous section, we will proceed with a simplified version of the equations to examine the fundamental characteristics. Let us consider the motion of a slowly rolling wheelset, that is, quasi-static motion. In this scenario, the inertia force of the wheelset can be regarded as negligible compared to the creep force. As a result, the equations of motion presented in Equation (2.5) can be simplified as follows:

$$\left.\begin{aligned} \frac{2\kappa_{22}}{v}\dot{y}_W - 2\kappa_{22}\psi_w = 0 \\ \frac{2\kappa_{11}b_0^2}{v}\dot{\psi}_w + \frac{2\kappa_{11}b_0\gamma}{r_0}y_w = 0 \end{aligned}\right\}. \tag{2.6}$$

This equation is a first-order, two-variable differential equation. From the upper expression in Equation (2.6), the following equation can be derived:

$$\dot{y}_W = v\psi_W. \tag{2.7}$$

By using this relation, Equation (2.6) can be reduced to a single equation:

$$\ddot{y}_{w} + \frac{v^2 \gamma}{b_0 r_0} y_{w} = 0. \tag{2.8}$$

The solution to the above equation is

$$y_{w} = A \sin\left(\sqrt{\frac{\gamma}{b_0 r_0}} vt\right). \tag{2.9}$$

In this case, the period T and kinematic hunting wavelength S_1 are expressed as follows:

$$T = \frac{2\pi}{v}\sqrt{\frac{b_0 r_0}{\gamma}}, \tag{2.10}$$

$$S_1 = Tv = 2\pi\sqrt{\frac{b_0 r_0}{\gamma}}. \tag{2.11}$$

Figure 2.5 illustrates the kinematic hunting motion. As indicated by the wavelength S_1, the larger the value of the tread gradient γ, the shorter the wavelength. The kinematic hunting motion is critically stable. However, due to the mass and inertia effects of the wheelset, the motion of a single wheelset tends to amplify over time. An actual eigenvalue analysis of Equation (2.5) confirms that the system remains unstable under all running velocities.

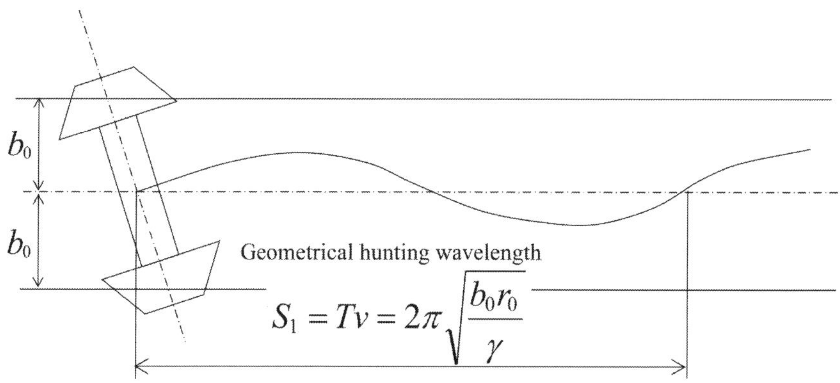

FIGURE 2.5 Geometrical hunting wavelength for the wheel axis.

2.2.3 Pure Rolling of a Wheelset

The wheelset generates a longitudinal creep force by lateral displacement, which results in a yaw moment. This effect offers a significant advantage by enabling the self-steering function during passage through a curve. Let us consider a wheelset running through a curve as shown in Figure 2.6. In a curve, the lengths of the high and low rails are different, and in order for the wheelset to roll smoothly on the curve, the wheels on the high and low rails must have radius differences that do not generate longitudinal creep forces. In this case, the circumferential velocity v_L at the wheel contact point on the low rail is

$$v_L = (R - b_0)\dot{\psi} = (r_0 - \gamma y_0)\Omega, \quad (2.12)$$

where Ω is the angular velocity of wheel rolling, $\dot{\psi}$ is the yaw angular velocity of the wheelset as viewed from the curve centre and y_0 is the wheelset lateral displacement with respect to the high rail. In contrast, the circumferential velocity v_R of the wheel contact point on the high rail is

$$v_R = (R + b_0)\dot{\psi} = (r_0 + \gamma y_0)\Omega. \quad (2.13)$$

As shown in Figure 2.6, if the wheel rolls without generating creep forces, that is, pure rolling, then

$$v_L : v_R = (R - b_0):(R + b_0). \quad (2.14)$$

From these relationships, the pure rolling displacement y_0 is

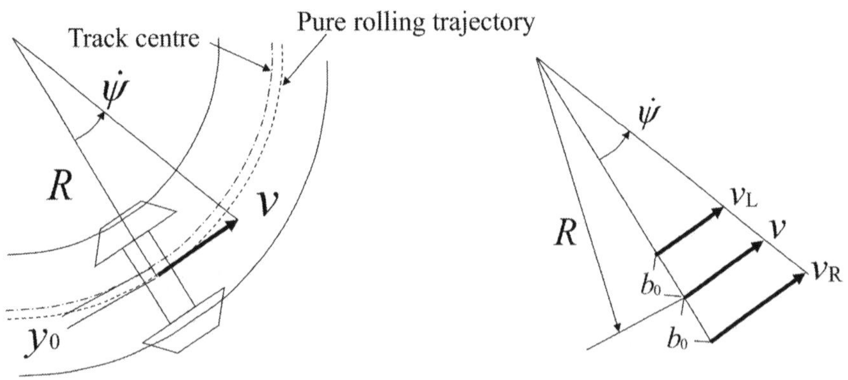

FIGURE 2.6 Wheelset during curve passage and illustration of pure rolling.

$$y_0 = \frac{b_0 r_0}{R\gamma}. \tag{2.15}$$

The pure rolling displacement represents the equilibrium point of the lateral motion of a wheelset running on the curve, and a longitudinal creep force is generated when the wheelset deviates from this equilibrium point. The locus of this pure rolling displacement is referred to as the pure rolling trajectory. From the perspective of wheel motion from a relative coordinate system based on the pure rolling trajectory, the equation of wheelset motion can be extended from the case of running on a straight line to the case of running on a curve.

Block diagrams can be used to explain the wheelset running on a curve as a relative coordinate motion with respect to a pure rolling trajectory. The external force acting on the wheelset is the centrifugal force while running on the curve. When a cant is introduced on a curve, the force in the lateral direction parallel to the track plane is referred to as the excessive centrifugal force and is expressed as

$$F_c = m_w \left(\frac{v^2}{R} - \frac{C}{G} g \right), \tag{2.16}$$

where R is the curve radius (m), C is the cant (m), G is the gauge (m) and g is the acceleration due to gravity (m/s^2). The wheelset has a flange to prevent deviation from the rail, and the lateral rigidity of the rail acts as a spring force against lateral displacement beyond the flange clearance δ. Figure 2.7 illustrates a block diagram of this mechanism.

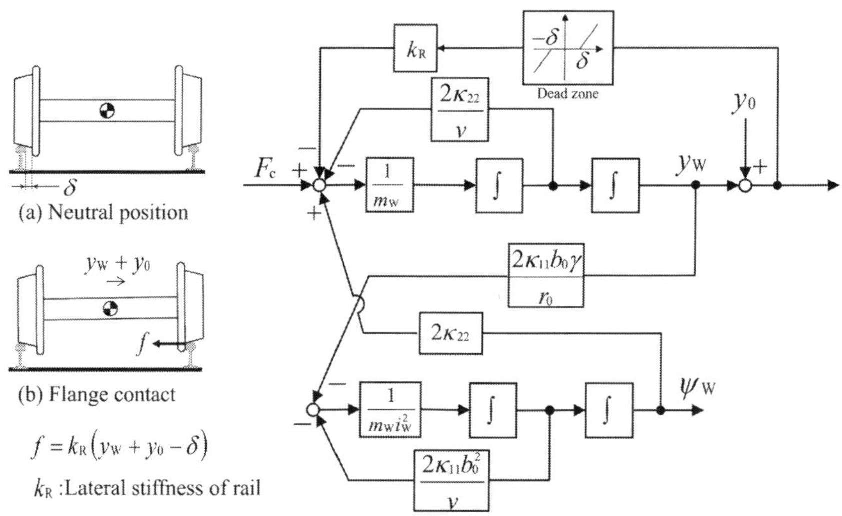

FIGURE 2.7 Wheelset when passing through a curve and block diagram of the rail's rigidity against lateral displacement.

2.3 DYNAMICS OF A TWO-AXLE BOGIE

The previous section focused on a single wheelset and described the characteristic dynamics of its hunting behaviour and self-steering function. In actual vehicle motion analysis, it is essential to extend the single wheelset model to a bogie or a single-vehicle model. In this section, the two-axle bogie shown in Figure 2.8 is used as a typical example for the construction of the equations of motion.

2.3.1 Configuration of the Bogie Model

In the case of a two-axle bogie, a 6-DOF model is used in the analysis, which accounts for the two sets of front and rear wheels, the lateral displacement of the bogie frame and the yaw angle as model DOFs when discussing the running stability. When the model has many DOFs, deriving the equations of motion from the equilibrium of the creep and spring forces becomes labour intensive. However, because the creep force between the wheel and the rail is not an energy conserving system, the Lagrange method cannot be directly applied. First, we consider a scenario where no creep force is generated between the wheels and the rails as if the bogie were riding on air. In this case, the kinetic energy of the front wheelset, rear wheelset and bogie frame of a two-axle bogie can be expressed as follows:

$$T = \frac{1}{2}m_W\dot{y}_{W1}^2 + \frac{1}{2}m_W i_W^2\dot{\psi}_{W1}^2 + \frac{1}{2}m_W\dot{y}_{W2}^2 + \frac{1}{2}m_W i_W^2\dot{\psi}_{W2}^2 + \frac{1}{2}m_T\dot{y}_T^2 + \frac{1}{2}m_T i_T^2\dot{\psi}_T^2, \tag{2.17}$$

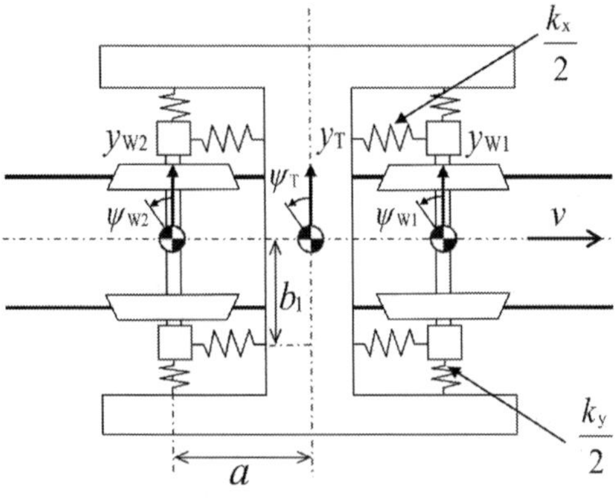

FIGURE 2.8 6-DOF biaxial bogie model.

where m_T is the bogie frame mass and i_T is the yaw radius of inertia around the centre of gravity of the bogie frame. The potential energy due to the axle spring is

$$\begin{aligned} U &= \frac{1}{2}k_x\left(b_1\psi_{W1}-b_1\psi_T\right)^2+\frac{1}{2}k_x\left(b_1\psi_{W2}-b_1\psi_T\right)^2 \\ &+\frac{1}{2}k_y\left(y_{W1}-y_T-a\psi_T\right)^2+\frac{1}{2}k_y\left(y_{W2}-y_T+a\psi_T\right)^2 \end{aligned}, \tag{2.18}$$

where a is half the distance between the bogie axles. From Equations (2.17) and (2.18), the Lagrangian $L = T - U$ is obtained, and the Lagrangian equation of motion is

$$\frac{d}{dt}\left(\frac{\partial L}{\partial \dot{X}}\right)-\frac{\partial L}{\partial X}=0. \tag{2.19}$$

Here, the generalized coordinate is defined as

$$\boldsymbol{X}=\begin{bmatrix} y_{W1} & \psi_{W1} & y_{W2} & \psi_{W2} & y_T & \psi_T \end{bmatrix}^T.$$

The following equations of motion are obtained by calculating Equation (2.19):

$$\boldsymbol{M}\ddot{\boldsymbol{X}}+\boldsymbol{K}\boldsymbol{X}=0, \tag{2.20}$$

where

$$\boldsymbol{M}=\operatorname{diag}\left(m_W, m_W i_W^2, m_W, m_W i_W^2, m_T, m_T i_T^2\right),$$

$$\boldsymbol{K}=\begin{bmatrix} k_y & 0 & 0 & 0 & -k_y & -k_y a \\ 0 & k_x b_1^2 & 0 & 0 & 0 & -k_x b_1^2 \\ 0 & 0 & k_y & 0 & -k_y & k_y a \\ 0 & 0 & 0 & k_x b_1^2 & 0 & -k_x b_1^2 \\ -k_y & 0 & -k_y & 0 & 2k_y & 0 \\ -k_y a & -k_x b_1^2 & k_y a & -k_x b_1^2 & 0 & 2k_x b_1^2+2k_y a^2 \end{bmatrix}.$$

Next, the terms due to the creep force between the wheel and the rail are considered. From the equation of motion for a single wheelset given in Equation (2.5), the damping matrix due to creep force $\boldsymbol{C}_c$ and the restoring force matrix due to creep force $\boldsymbol{K}_c$ are as follows:

$$\boldsymbol{C}_c = \frac{1}{v}\begin{bmatrix} 2\kappa_{22} & 0 & 0 & 0 & 0 & 0 \\ 0 & 2\kappa_{11}b_0^2 & 0 & 0 & 0 & 0 \\ 0 & 0 & 2\kappa_{22} & 0 & 0 & 0 \\ 0 & 0 & 0 & 2\kappa_{11}b_0^2 & 0 & 0 \\ 0 & 0 & 0 & 0 & 0 & 0 \\ 0 & 0 & 0 & 0 & 0 & 0 \end{bmatrix},$$

$$\boldsymbol{K}_c = \begin{bmatrix} 0 & -2\kappa_{22} & 0 & 0 & 0 & 0 \\ \dfrac{2\kappa_{11}b_0\gamma}{r_0} & 0 & 0 & 0 & 0 & 0 \\ 0 & 0 & 0 & -2\kappa_{22} & 0 & 0 \\ 0 & 0 & \dfrac{2\kappa_{11}b_0\gamma}{r_0} & 0 & 0 & 0 \\ 0 & 0 & 0 & 0 & 0 & 0 \\ 0 & 0 & 0 & 0 & 0 & 0 \end{bmatrix}.$$

After these matrices are combined with Equation (2.20), the following equation can be obtained:

$$\boldsymbol{M}\ddot{\boldsymbol{X}} + \boldsymbol{C}_c\dot{\boldsymbol{X}} + \left(\boldsymbol{K} + \boldsymbol{K}_c\right)\boldsymbol{X} = \boldsymbol{0}. \tag{2.21}$$

As mentioned above, the equations of motion can be easily constructed by separating the terms of the creep force and suspension forces.

2.3.2 Bogie on a Curve

When a bogie runs on a curve, it is crucial to consider how the track's curve radius is incorporated into the equations of motion, as well as how the stiffness of the axle supports influences the wheelset's motion and the bogie's attitude. In this case, the bogie motion can be described by using a relative coordinate system along the curved track. For a wheelset alone, the motion can be similarly characterised as in the case of a straight track, where the pure rolling displacement serves as the reference equilibrium point for lateral displacement. However, in the case of a bogie, it is necessary to incorporate a coordinate system for the yaw angles of both the front and rear wheelsets. As illustrated in Figure 2.9(a), the equilibrium points for the yaw angles of the front and rear axles are located in the direction normal to the track for each wheelset. In this scenario, the yaw angles of the front and rear wheelsets, as viewed with respect to the centre of gravity of the bogie frame, are expressed as follows:

$$\left.\begin{aligned} \psi_{W_1} &\to \psi_{W_1} - \frac{a}{R} \\ \psi_{W_2} &\to \psi_{W_2} + \frac{a}{R} \end{aligned}\right\}, \tag{2.22}$$

and this results in a spring force acting through the bogie frame,

$$\boldsymbol{K}\begin{bmatrix} y_{W1} \\ \psi_{W1} - \dfrac{a}{R} \\ y_{W2} \\ \psi_{W2} + \dfrac{a}{R} \\ y_T \\ \psi_T \end{bmatrix} = \boldsymbol{KX} - \boldsymbol{K}\begin{bmatrix} 0 \\ \dfrac{a}{R} \\ 0 \\ -\dfrac{a}{R} \\ 0 \\ 0 \end{bmatrix}, \tag{2.23}$$

and an external force term inversely proportional to the curve radius R is added to the equation of motion in Equation (2.21).

Next, the quasi-static posture of the wheelset and bogie frame on the curve is described. In the case of a circular curve, the acceleration and velocity vectors on the left-hand side of the equation of motion in Equation (2.21) are assumed to be zero, resulting in the following force balance equation:

$$\left(\boldsymbol{K} + \boldsymbol{K}_c\right)\boldsymbol{X} = \boldsymbol{K}\boldsymbol{\psi}_\rho + \boldsymbol{F}_c, \tag{2.24}$$

where $\boldsymbol{\psi}_\rho$ is obtained from Equation (2.23) as

$$\boldsymbol{\psi}_\rho = \frac{a}{R}\begin{bmatrix} 0 & 1 & 0 & -1 & 0 & 0 \end{bmatrix}^T. \tag{2.25}$$

$\boldsymbol{F}_c$ is the term for the excessive centrifugal force acting on the wheelset and bogie frame and is described as

$$F_c = \left(\frac{v^2}{R} - \frac{C}{G}g\right)\begin{bmatrix} m_W & 0 & m_W & 0 & m_T & 0 \end{bmatrix}^T. \tag{2.26}$$

Then the quasi-static state $\boldsymbol{X}$ from Equation (2.24) can be obtained as

$$X = \left(\boldsymbol{K} + \boldsymbol{K}_c\right)^{-1}\left(\boldsymbol{K}\boldsymbol{\psi}_\rho + \boldsymbol{F}_c\right). \tag{2.27}$$

From the equation above, the lateral displacement and yaw angle on a curve can be calculated for both wheelsets and the bogie frame. The yaw angle of the wheelset corresponds to the angle of attack, which is the angle between the direction of wheel travel and the tangent direction of the track. A larger angle of attack results in an increase in lateral force. It is important to note that the lateral displacement of the wheelset is referenced from the pure rolling displacement, rather than the displacement from the track centre, as illustrated in Figure 2.9.

On sharp curves, flange contact occurs due to lateral displacement of the bogie front wheelset. Although Equation (2.27) is a simplified, linearised formula, it is useful when discussing the steering performance of bogies. Figure 2.10 shows the relationship between the lateral displacement of wheelset and the flange contact considering slack S and flange clearance δ. As the value of tread gradient γ increases, the wheelset can avoid flange contact even on a sharp curve. Figure 2.11 shows the results of lateral displacement of the leading wheelset derived from Equation (2.27). In this case, the effect of centrifugal force $\boldsymbol{F}_c$ is omitted for simplicity. Due to the restriction by the bogie frame and primary suspension stiffness, the lateral displacement of the leading wheelset of the bogie increases. In this case, the tread gradient γ plays an important role in avoiding flange contact.

The considerations outlined in this section are for linear systems. In practice, however, as the wheel diameter difference and angle of attack increases, the

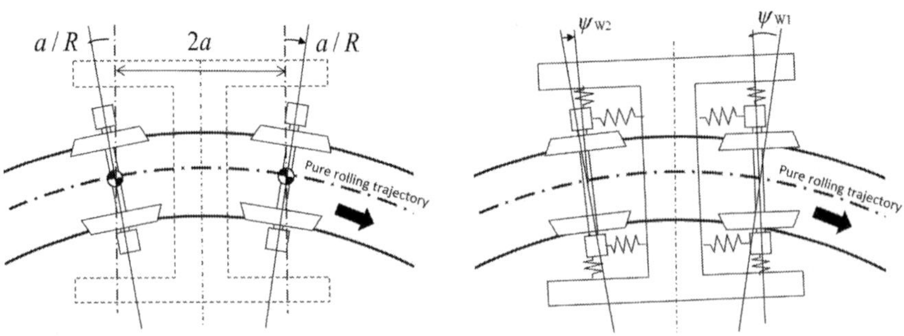

(a) Position of the equilibrium point (b) Definition of wheel yaw angle

FIGURE 2.9 Coordinate system and bogie posture when passing through a curve.

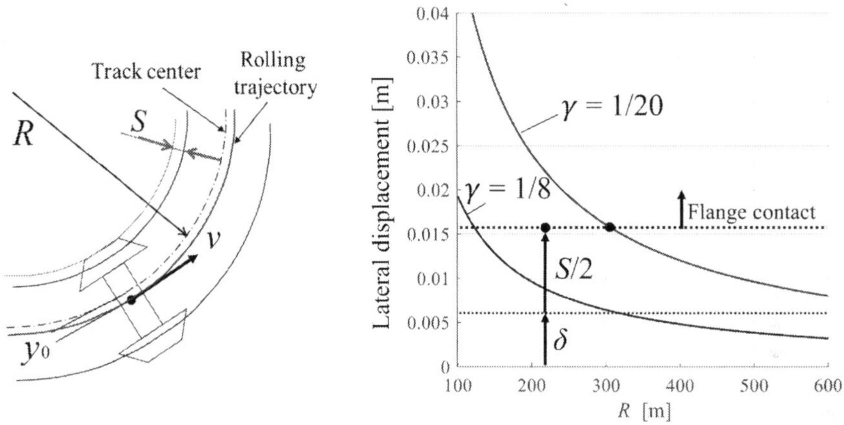

FIGURE 2.10 Relation between lateral displacement of wheelset and flange contact considering slack S (= 20 mm) and flange clearance δ (= 6 mm).

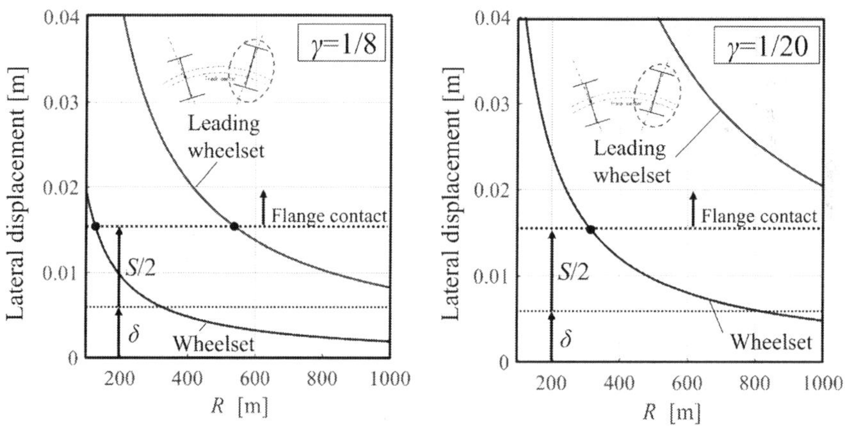

FIGURE 2.11 Relation between lateral displacement of wheelset with bogie frame restriction and flange contact considering slack and flange clearance.

creep ratio increases, causing saturation of the creep force and non-negligible spin moments. Therefore, the theory based on a linear creep force, which is not applicable to the negotiation of sharp curves, should be replaced by numerical analysis using multi-body dynamics, as discussed in Chapter 3, for quantitative analysis of such curves. Although Equation (2.27) is a simplified, linearised formula, it is still useful when discussing the steering performance of bogies including those with unconventional designs [11–18].

2.3.3 Geometric Hunting Motion of Bogies

To facilitate a deeper understanding of bogie dynamics, a simplified bogie model is used to explain the characteristics of bogie hunting behaviour. For the sake of simplicity, a bogie model with rigid spring stiffness is used. By assuming the primary suspension stiffness to be infinite, the lateral displacement of front axis y_{W1} and yaw angle ψ_{W_1} are as follows.

$$\left.\begin{aligned} y_{W_1} &= y_T + a\psi_T \\ \psi_{W_1} &= \psi_T \end{aligned}\right\}, \tag{2.28}$$

where y_T is the lateral displacement of the bogie frame centre of gravity and ψ_T is the yaw angle of the bogie frame.

Lateral displacement of the rear wheelset y_{W2} and yaw angle ψ_{w2} are denoted as follows:

$$\left.\begin{aligned} y_{W2} &= y_T - a\psi_T \\ \psi_{w2} &= \psi_T \end{aligned}\right\}. \tag{2.29}$$

By calculating the combined creep forces acting on the four wheels in the lateral direction and the moment around the centre of the bogie frame, the equations of motion for a rigid bogie can be described as follows:

$$\left.\begin{aligned} m_T\ddot{y}_T + \frac{4\kappa}{v}\dot{y}_T - 4\kappa\psi_T = 0 \\ I_T\ddot{\psi}_T + \frac{4\left(a^2 + b_0^2\right)\kappa}{v}\dot{\psi}_T + \frac{4\kappa b_0\gamma}{r_0}y_T = 0 \end{aligned}\right\}, \tag{2.30}$$

where m_T is the mass of the bogie (bogie frame and wheelset) and I_T is the moment of inertia around the centre of gravity of the bogie.

This equation of motion is similar to that for a single wheelset as described in Equation (2.5). Assuming quasi-static motion and neglecting the inertia term, the following equation can be derived:

$$\left.\begin{aligned} \frac{4\kappa}{v}\dot{y}_T - 4\kappa\psi_T = 0 \\ \frac{4\left(a^2 + b_0^2\right)\kappa}{v}\dot{\psi}_T + \frac{4\kappa b_0\gamma}{r_0}y_T = 0 \end{aligned}\right\}. \tag{2.31}$$

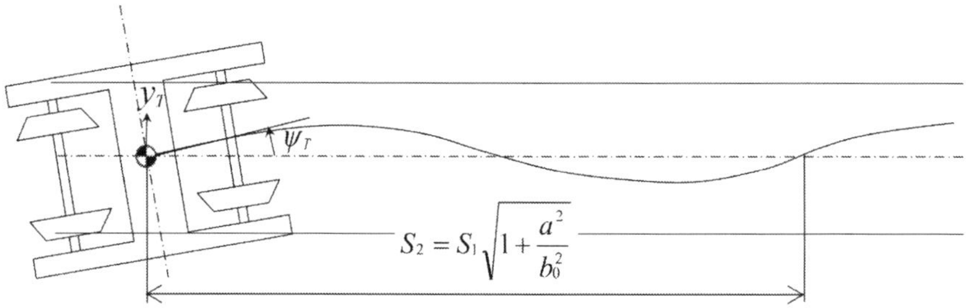

FIGURE 2.12 Geometric hunting behaviour of a rigid bogie.

From this equation, the lateral displacement equation is derived as follows:

$$\ddot{y}_{\mathrm{T}} + \frac{v^2\gamma_0}{\left(a^2+b_0^2\right)r_0} y_{\mathrm{T}} = 0 \,. \tag{2.32}$$

The geometrical hunting wavelength of the bogie is as follows:

$$S_2 = S_1\sqrt{1+\frac{a^2}{b_0^2}} \,. \tag{2.33}$$

As indicated by this equation, the geometric hunting wavelength S_1 for a single wheelset is related to the geometric hunting wavelength S_2 for the bogie. It can also be found that, for example, an increase in the wheelbase results in an increase in the hunting wavelength, which leads to more stable bogie characteristics. Although these results are derived under the assumption of neglecting inertia term, similar qualitative trends are typically observed even when more complex models are employed (see Figure 2.12).

2.4 VEHICLE MODEL

2.4.1 Vehicle Model Example for Vertical Motion (6 DOFs)

A vehicle model representing the vertical motion of railway vehicles is used for the analysis of riding comfort or to estimate the vertical track irregularities, as illustrated in Figure 2.13.

The model is a single-vehicle linear model that is assumed to run on a straight track, and has 6 DOFs, that is, 2 DOFs (vertical movement and pitch) for the car body and 2 DOFs (vertical movement and pitch) for each of the two bogies.

In the model, Z_c represents the vertical movement of the car body, Z_{t1} represents the vertical movement of the front bogie, Z_{t2} represents the vertical movement of

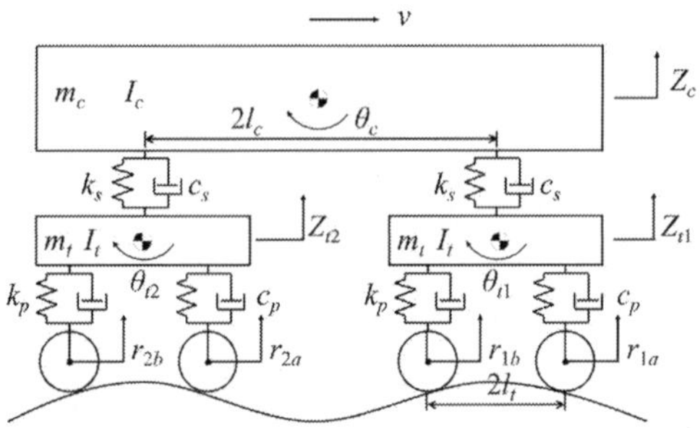

FIGURE 2.13 6-DOF vehicle model.

the rear bogie, θ_c represents the pitch angle of the car body, θ_{t1} represents the pitch angle of the front bogie and θ_{t2} represents the pitch angle of the rear bogie. Inputs r_{1a}, r_{1b}, r_{2a} and r_{1b} are the forced inputs due to vertical track displacement given to each wheelset.

The equations of motion when travelling on a straight track are expressed as follows:

Car body bounce:

$$\ddot{Z}_c = \frac{1}{m_c}\left[c_s\dot{Z}_{t1} + c_s\dot{Z}_{t2} - 2c_s\dot{Z}_c + k_s Z_{t1} + k_s Z_{t2} - 2k_s Z_c\right]. \tag{2.34}$$

Car body pitch:

$$\ddot{\theta}_c = \frac{1}{I_c}\left[-c_s l_c\dot{Z}_{t1} + c_s l_c\dot{Z}_{t2} - 2c_s l_c^2\dot{\theta}_c - k_s l_c Z_{t1} + k_s l_c Z_{t2} - 2k_s l_c^2\theta_c\right]. \tag{2.35}$$

Front bogie bounce:

$$\begin{aligned}\ddot{Z}_{t1} = \frac{1}{m_t}\Big[&\left(-c_s - 2c_p\right)\dot{Z}_{t1} + c_s\dot{Z}_c + c_s l_c\dot{\theta}_c \\ &+\left(-k_s - 2k_p\right)Z_{t1} + k_s Z_c + k_s l_c\theta_c + c_p\dot{r}_{1a} + c_p\dot{r}_{1b} + k_p r_{1a} + k_p r_{1b}\Big].\end{aligned} \tag{2.36}$$

Front bogie pitch:

$$\ddot{\theta}_{t1} = \frac{1}{I_t}\left[-c_p l_t\dot{r}_{1a} + c_p l_t\dot{r}_{1b} - 2c_p l_t^2\dot{\theta}_{t1} - k_p l_t r_{1a} + k_p l_t r_{1b} - 2k_p l_t^2\theta_{t1}\right]. \tag{2.37}$$

Rear bogie bounce:

$$\ddot{Z}_{t2} = \frac{1}{m_t}\Big[\left(-c_s - 2c_p\right)\dot{Z}_{t2} + c_s\dot{Z}_c - c_s l_c\dot{\theta}_c + \left(-k_s - 2k_p\right)Z_{t2} + k_s Z_c - k_s l_c\theta_c + c_p\dot{r}_{2a} + c_p\dot{r}_{2b} + k_p r_{2a} + k_p r_{2b}\Big]. \tag{2.38}$$

Rear bogie pitch:

$$\ddot{\theta}_{t2} = \frac{1}{I_t}\left[-c_p l_t\dot{r}_{2a} + c_p l_t\dot{r}_{2b} - 2c_p l_t^2\dot{\theta}_{t2} - k_p l_t r_{2a} + k_p l_t r_{2b} - 2k_p l_t^2\theta_{t2}\right]. \tag{2.39}$$

Equations (2.34)–(2.39) can be expressed as an MCK-type equation as

$$\boldsymbol{M}\ddot{\boldsymbol{Z}} + \boldsymbol{C}\dot{\boldsymbol{Z}} + \boldsymbol{K}\boldsymbol{Z} = \boldsymbol{D}\dot{\boldsymbol{r}} + \boldsymbol{E}\boldsymbol{r}, \tag{2.40}$$

where $\boldsymbol{M}$, $\boldsymbol{C}$ and $\boldsymbol{K}$ are the mass, damping and stiffness matrices, respectively, and $\ddot{\boldsymbol{Z}}$, $\dot{\boldsymbol{Z}}$, $\boldsymbol{Z}$, $\dot{\boldsymbol{r}}$ and $\boldsymbol{r}$ are the acceleration vector, velocity vector, displacement vector, derivative of the external force vector and external force vector, respectively. Each matrix in the above equation is given below.

$$\boldsymbol{M} = \mathrm{diag}\left(m_c, \frac{I_c}{l_c^2}, m_t, \frac{I_t}{l_t^2}, m_t, \frac{I_t}{l_t^2}\right),$$

$$\boldsymbol{C} = \begin{bmatrix} 2c_s & 0 & -c_s & 0 & -c_s & 0 \\ 0 & 2c_s & -c_s & 0 & c_s & 0 \\ -c_s & -c_s & 2c_p + c_s & 0 & 0 & 0 \\ 0 & 0 & 0 & 2c_p & 0 & 0 \\ -c_s & c_s & 0 & 0 & 2c_p + c_s & 0 \\ 0 & 0 & 0 & 0 & 0 & 2c_p \end{bmatrix},$$

$$
\boldsymbol{K} = \begin{bmatrix} 2k_s & 0 & -k_s & 0 & -k_s & 0 \\ 0 & 2k_s & -k_s & 0 & k_s & 0 \\ -k_s & -k_s & 2k_p + k_s & 0 & 0 & 0 \\ 0 & 0 & 0 & 2k_p & 0 & 0 \\ -k_s & k_s & 0 & 0 & 2k_p + k_s & 0 \\ 0 & 0 & 0 & 0 & 0 & 2k_p \end{bmatrix},
$$

$$
\boldsymbol{D} = \begin{bmatrix} 0 & 0 & 0 & 0 \\ 0 & 0 & 0 & 0 \\ c_p & c_p & 0 & 0 \\ c_p & -c_p & 0 & 0 \\ 0 & 0 & c_p & c_p \\ 0 & 0 & c_p & -c_p \end{bmatrix},
$$

$$
\boldsymbol{E} = \begin{bmatrix} 0 & 0 & 0 & 0 \\ 0 & 0 & 0 & 0 \\ k_p & k_p & 0 & 0 \\ k_p & -k_p & 0 & 0 \\ 0 & 0 & k_p & k_p \\ 0 & 0 & k_p & -k_p \end{bmatrix},
$$

$$
\boldsymbol{Z} = \begin{bmatrix} Z_c & l_c\theta_c & Z_{t1} & l_t\theta_{t1} & Z_{t2} & l_t\theta_{t2} \end{bmatrix}^T \text{ and } \boldsymbol{r} = \begin{bmatrix} r_{1a} & r_{1b} & r_{2a} & r_{2b} \end{bmatrix}^T.
$$

Time series analysis problems, such as vehicle dynamics simulation, require time integration of the equations of motion. This method is based on differential equations and is used to obtain the time history response of the vehicle through a large number of iterations. Common numerical integration methods include the Euler and Runge–Kutta.

2.4.2 Vehicle Model (17 DOFs)

The 17-DOF model depicted in Figure 2.14 is considered when discussing the running stability of a vehicle, which includes the 3 DOFs of lateral, yaw

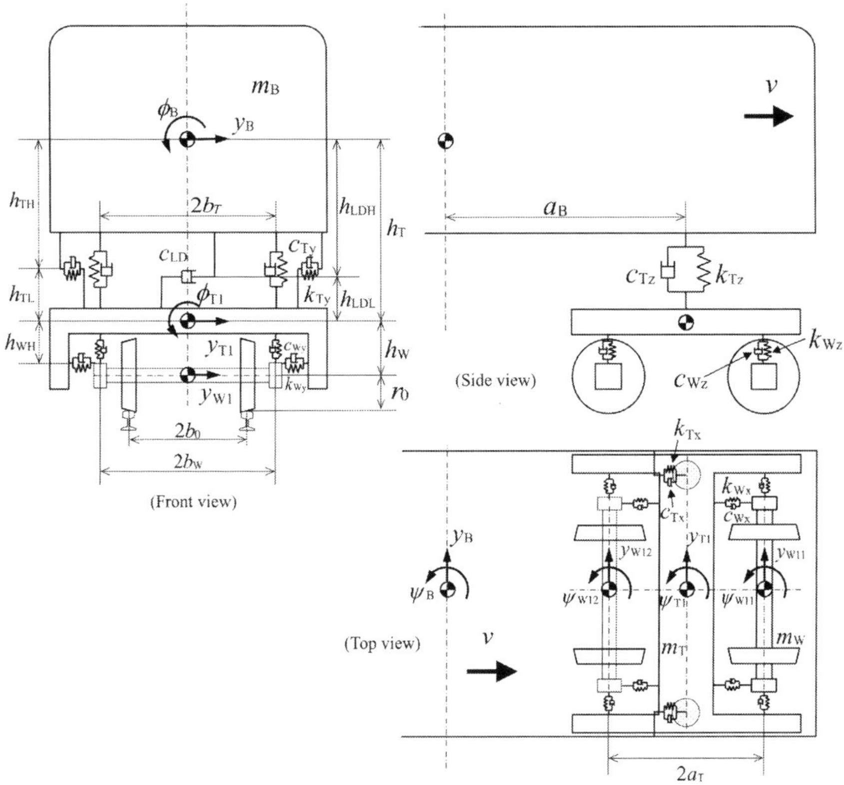

FIGURE 2.14 Vehicle model for hunting behaviour analysis.

and roll of the car body in addition to the model DOFs of the bogie as the degrees of motion of a single vehicle. A detailed description of these DOFs is provided in Table 2.1.

This motion DOF framework is intended to analyse the stability of the hunting behaviour of the vehicle. For example, to evaluate the ride quality against the high and low displacement of the track, the vertical movement, including pitching of the car body and bogie, should be considered.
The subscripts W, T and B indicate the wheelset, bogie frame and car body, respectively.

$i = 1$ indicates the front bogie, $i = 2$ indicates the rear bogie, $j = 1$ indicates the front axle of the bogie and $j = 2$ indicates the rear axle of the bogie. Even when the number of DOFs of the model increases, the approach to deriving the equations of motion is the same as that for the bogie model, and it is efficient to consider the

TABLE 2.1
Degrees of freedom per rigid body for a one-vehicle 17-DOF model

Component	Degree of freedom			Number of bodies	Number of DOFs
Car body	Lateral: y_{B}	Roll: ϕ_{B}	Yaw: ψ_{B}	1	3
Bogie frame	Lateral: $y_{\mathrm{T}i}$	Roll: $\phi_{\mathrm{T}i}$	Yaw: $\psi_{\mathrm{T}i}$	2	6
Wheelset	Lateral: $y_{\mathrm{W}_{ij}}$		Yaw: $\psi_{\mathrm{W}_{ij}}$	4	8
				Total	17

creep force and the stiffness and damping matrices due to the springs and dampers separately. The potential energy of the primary spring of the bogie is as follows:

$$\begin{aligned} U_{\mathrm{T}i} &= k_{\mathrm{W}x} b_{\mathrm{W}}^2 \left(\psi_{\mathrm{T}i} - \psi_{\mathrm{W}i1}\right)^2 + k_{\mathrm{W}x} b_{\mathrm{W}}^2 \left(\psi_{\mathrm{T}i} - \psi_{\mathrm{W}i2}\right)^2 \\ &+ k_{\mathrm{W}y}\left(y_{\mathrm{T}i} + a_{\mathrm{T}}\psi_{\mathrm{T}i} + h_{\mathrm{WH}}\phi_{\mathrm{T}i} - y_{\mathrm{W}i1}\right)^2 \\ &+ k_{\mathrm{W}y}\left(y_{\mathrm{T}i} + a_{\mathrm{T}}\psi_{\mathrm{T}i} + h_{\mathrm{WH}}\phi_{\mathrm{T}i} - y_{\mathrm{W}i2}\right)^2 \\ &+ k_{\mathrm{W}z} b_{\mathrm{W}}^2 \left(\phi_{\mathrm{T}i} - \phi_{\mathrm{W}i1}\right)^2 + k_{\mathrm{W}z} b_{\mathrm{W}}^2 \left(\phi_{\mathrm{T}i} - \phi_{\mathrm{W}i2}\right)^2. \end{aligned} \tag{2.41}$$

The roll angle of the wheelset can be approximated using the tread gradient γ as follows:

$$\phi_{\mathrm{W}ij} = \frac{\gamma}{b_0} y_{\mathrm{W}i1}. \tag{2.42}$$

The potential energy of the springs in the secondary suspension system between the two bogies and the car body is calculated as

$$\begin{aligned} U_2 &= k_{\mathrm{T}x} b_{\mathrm{T}}^2 \left(\psi_{\mathrm{B}} - \psi_{\mathrm{T}1}\right)^2 + k_{\mathrm{T}x} b_{\mathrm{T}}^2 \left(\psi_{\mathrm{B}} - \psi_{\mathrm{T}2}\right)^2 \\ &+ k_{\mathrm{T}y}\left(y_{\mathrm{B}} + h_{\mathrm{TH}}\phi_{\mathrm{B}} + a_{\mathrm{B}}\psi_{\mathrm{B}} - y_{\mathrm{T}1} + h_{\mathrm{TL}}\phi_{\mathrm{T}1}\right)^2 \\ &+ k_{\mathrm{T}y}\left(y_{\mathrm{B}} + h_{\mathrm{TH}}\phi_{\mathrm{B}} - a_{\mathrm{B}}\psi_{\mathrm{B}} - y_{\mathrm{T}2} + h_{\mathrm{TL}}\phi_{\mathrm{T}2}\right)^2 \\ &+ k_{\mathrm{T}z} b_{\mathrm{T}}^2 \left(\phi_{\mathrm{B}} - \phi_{\mathrm{T}1}\right)^2 + k_{\mathrm{T}z} b_{\mathrm{T}}^2 \left(\phi_{\mathrm{B}} - \phi_{\mathrm{T}2}\right)^2. \end{aligned} \tag{2.43}$$

With regard to the damping force of the primary and secondary dampers, the potential energy of the springs shown above can be used as a reference, the spring constant can be replaced with the damping coefficient and the displacement and angle can be converted to velocity and angular velocity, respectively, to define the dissipation function. For this reason, we omit a description of the dissipation function of the damper in parallel with the spring. The dissipation function for the lateral motion damper (two in total, one on the front bogie and one on the rear bogie), which is not in parallel to the spring, is as follows:

$$\begin{aligned} D_{\mathrm{LD}} &= \frac{1}{2} c_{\mathrm{LD}} \left(\dot{y}_{\mathrm{B}} + h_{\mathrm{LDH}} \dot{\phi}_{\mathrm{B}} + a_{\mathrm{B}} \dot{\psi}_{\mathrm{B}} - \dot{y}_{\mathrm{T1}} + h_{\mathrm{LDL}} \dot{\phi}_{\mathrm{T1}} \right)^2 \\ &+ \frac{1}{2} c_{\mathrm{LD}} \left(\dot{y}_{\mathrm{B}} + h_{\mathrm{LDH}} \dot{\phi}_{\mathrm{B}} - a_{\mathrm{B}} \dot{\psi}_{\mathrm{B}} - \dot{y}_{\mathrm{T2}} + h_{\mathrm{LDL}} \dot{\phi}_{\mathrm{T2}} \right)^2 . \end{aligned} \tag{2.44}$$

From the energy defined above, the damping and stiffness matrices for the support mechanism can be calculated and combining them with the creep force term according to the procedure in the previous section allows us to derive the equations of motion.

Eigenvalue analysis is conducted for the equations of motion of the single-vehicle model derived in the previous section. The equation of motion, which incorporates the spring force, damping force, and creep force resulting from the support mechanism described earlier, is given by:

$$\boldsymbol{M}\ddot{\boldsymbol{X}} + \boldsymbol{C}\dot{\boldsymbol{X}} + \boldsymbol{K}\boldsymbol{X} = \boldsymbol{0}. \tag{2.45}$$

The corresponding state equation is

$$\frac{d}{dt}\begin{bmatrix} \dot{\boldsymbol{X}} \\ \boldsymbol{X} \end{bmatrix} = \begin{bmatrix} -\boldsymbol{M}^{-1}\boldsymbol{C} & -\boldsymbol{M}^{-1}\boldsymbol{K} \\ \boldsymbol{I} & \boldsymbol{0} \end{bmatrix} \begin{bmatrix} \dot{\boldsymbol{X}} \\ \boldsymbol{X} \end{bmatrix}. \tag{2.46}$$

Where $\boldsymbol{I}$ and 0 are the identity matrix and zero matrix, respectively, which are 17×17 square matrices for the 17-DOF vehicle motion model. In this case, the matrix

$\boldsymbol{A}$ is defined as

$$\boldsymbol{A} \equiv \begin{bmatrix} -\boldsymbol{M}^{-1}\boldsymbol{C} & -\boldsymbol{M}^{-1}\boldsymbol{K} \\ \boldsymbol{I} & \boldsymbol{0} \end{bmatrix}. \tag{2.47}$$

The characteristic equation used to determine stability is then given by

$$\det\left(s\boldsymbol{I} - \boldsymbol{A}\right) = 0. \tag{2.48}$$

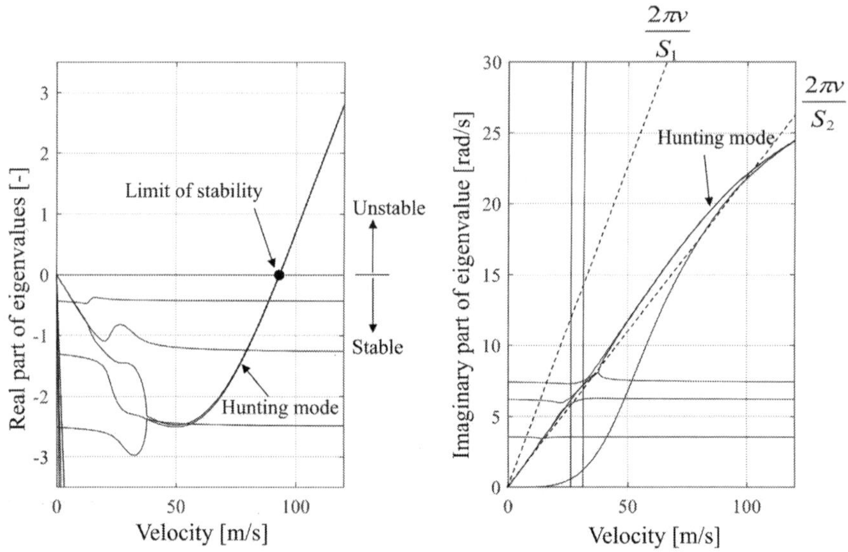

FIGURE 2.15 Eigenvalue analysis of a vehicle model.

Figure 2.15 presents the results of an eigenvalue analysis with running velocity as a parameter. As illustrated in the figure, the real part of the eigenvalues indicates that there is a mode that becomes unstable at a running velocity of approximately 93 m/s. This velocity is referred to as the critical hunting velocity. The imaginary part of the eigenvalues indicates that the value of this mode increases with increasing running velocity. The imaginary part of the eigenvalue of this destabilizing mode shows that the value increases as the running velocity increases. This mode is identified as the hunting mode, with the imaginary part of the eigenvalue indicating that it exists between the angular velocities derived from the geometric hunting behaviour of the wheelset and those of the rigid bogie.

The velocity at which the real part of the eigenvalues becomes positive can be altered by modifying various vehicle parameters, particularly the longitudinal and lateral support stiffness of the axlebox. To increase the critical hunting velocity, effective measures include reducing the conicity, increasing the wheelbase, decreasing the mass of the wheelset and bogie, and enhancing yaw damping of the bogie by adding yaw dampers.

REFERENCES

1. J.J. Kalker, Survey of wheel/rail rolling contact theory, *Vehicle System Dynamics*, 8(4), 317–358, 1979.
2. Z. Shen, K.J. Hedrick, J.A. Elkins, A comparison of alternative creep force models for rail vehicle dynamic analysis, *Proceedings of 8th IAVSD Symposium*, Taylor & Francis, 591–605, 1983.

3. J.J. Kalker, *Three-Dimensional Elastic Bodies in Rolling Contact, Solid Mechanics and Its Applications*, Vol. 2, Kluwer Academic, 1990.
4. J. Goree, A.H. Law, User's manual for Kalker's simplified nonlinear creep theory, *Department of Transportation Report*, No. FRA/ORD/78/06, 1977.
5. K.L. Johnson, *Contact Mechanics*, Cambridge University Press, 1987.
6. A.H. Wickens, Office of research and experiment (ORE) of the International Union of Railways (UIC), *Question C116*, Report 4, 1974.
7. T. Matsudaira, Hunting problem of high-speed railway vehicles with special reference to bogie design for the New Tokaido Line, *Proceedings of Institution Mechanical Engineering*, 180, 58–66, 1965.
8. A.H. Wickens, The dynamic stability of railway vehicle wheelsets and bogies having profiled wheels, *International Journal of Solids and Structures*, 1, 319–341, 1965.
9. A.O. Gilchrist, A.E.W. Hobbs, B.L. King, V. Washby, The riding of two particular designs of four wheeled vehicle, *Proceedings of Institution Mechanical Engineering*, 180, 99–113, 1965.
10. R.A. Pooley, Assessment of the critical speeds of various types of four-wheeled vehicles, *British Railways Research Department Report E557*, 1965.
11. H. Scheffel, The hunting stability and curving ability of railway vehicles, *Rail International*, 5, 154–176, 1974.
12. H. Scheffel, A new design approach for railway vehicle suspension, *Rail International*, 5, 638–651, 1974.
13. A.H. Wickens, Steering and dynamic stability of railway vehicles, *Vehicle System Dynamics*, 5, 15–46, 1975.
14. A.H. Wickens, Stability criteria for articulated railway vehicles possessing perfect steering, *Vehicle System Dynamics*, 7, 33–48, 1979.
15. H. Scheffel, R.D. Fröhling, P.S. Heyns, Curving and stability of self-steering bogies having a variable constraint, *Vehicle System Dynamics*, Supplement 1, 217–232, 1994.
16. H. Scheffel, Unconventional bogie designs -Their practical basis and historical background, *Vehicle System Dynamics*, 24, 497–524, 1995.
17. Y. Suda, R.J. Anderson, K. Yamada, Dynamics of unsymmetric suspension trucks with semiactive control, *Vehicle System Dynamics*, 23, Supplement 1, 480–496, 1994.
18. Y. Suda, Y. Michitsuji, Improved curving performance using unconventional wheelset guidance design and wheel-rail interface–present and future solutions, *Vehicle System Dynamics*, 61(7), 881–1915, 2023.

3 Application of Vehicle Simulation to Railway Safety

3.1 MBD SIMULATION OF A RAILWAY VEHICLE

Typical objectives of vehicle motion simulation include analysis of high-speed running stability in a straight line, analysis of curve passing performance, vehicle response to track displacement on a straight or curved track, verification of running safety, derailment phenomena and overturning of vehicles. For these problems, it is sometimes possible to understand the phenomena using relatively simple models as described in Chapter 2. However, when a more detailed and quantitative discussion is required, modelling and analysis based on Multi-Body Dynamics (MBD) may be required [1]. By carrying out an analysis based on such modelling, it is possible to understand how each parameter of the vehicle influences the results, which can then be applied to vehicle development, design and evaluation.

In describing the equations of motion, it is first necessary to determine the coordinate system. Two typical coordinate systems are shown in Figure 3.1. The equations of motion can be obtained relatively straightforwardly in the vehicle movement coordinate system. As shown in Figure 3.1(a), a coordinate system is defined at the centre of the vehicle, which moves as the vehicle travels. By introducing this coordinate system, the DOFs for the longitudinal direction of each rigid body can be omitted if the inertial forces due to acceleration and deceleration and the longitudinal forces (for example, due to couplers) can be ignored. As described in Section 2.3.2 of Chapter 2, to analyse curving performance using the vehicle movement coordinate system, the centrifugal force generated when the vehicle travels along a curve is given to each rigid body as an acting force, while the origin of the longitudinal and yawing directions are given along the track under each wheelset.

In a typical hunting motion analysis, the focus is on the lateral movement and yawing of the vehicle. The translational DOFs of all rigid bodies are neglected in the vehicle moving coordinate system. Under the assumption that the vehicle runs at a constant speed, the angular velocity of rotation of the wheelset is assumed to be constant and the pitching of the wheelset is omitted. Furthermore, when hunting behaviour on a straight track is analysed, the wheelset rolling becomes a dependent

DOI: 10.1201/9781003362135-3

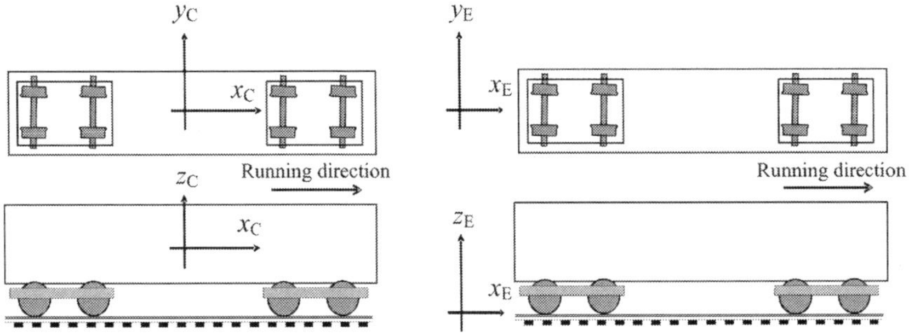

(a) Vehicle moving coordinate system (b) Fixed ground-based system

FIGURE 3.1 Coordinate system used in vehicle motion analysis.

variable of the lateral displacement by expressing the wheel tread profile. These reduce the DOFs of the wheelset to two: lateral and yaw. As a result, a single vehicle can be represented in motion with the 17 DOFs described in Section 2.4.2 in Chapter 2.

However, when the origin of the coordinate system is fixed to a point on the ground and the motion from that origin is analysed, the fixed ground-based coordinate system shown in Figure 3.1(b) is used. To describe the motion in a fixed ground-based coordinate system, the equations of motion are solved with position and rotation transformations of the vehicle attitude based on absolute coordinates. The numerical solution of the equations of motion is therefore more difficult to obtain than in the case of a vehicle moving coordinate system. However, in the equations of motion for the vehicle moving coordinate system, it is necessary to add terms for yawing of the vehicle body and excessive centrifugal force as an external force when passing through a curve, whereas this operation is not necessary when using a fixed ground-based coordinate system. The Coriolis forces acting on the vehicle are also taken into account from the outset in a fixed ground-based coordinate system. To construct a single-vehicle model in a fixed ground-based coordinate system, the seven rigid bodies (one body, two bogie frames and four wheelsets) comprising the vehicle are given 6 DOFs in translation and rotation, giving an equation of motion with 42 DOFs. It is assumed that most recent commercial vehicle motion analysis programmes also define 42 DOFs. For reference, Table 3.1 lists examples of rail vehicle motion analysis programmes [2].

3.2 CURVING SIMULATION WITH MBD TOOL

3.2.1 Simulation Procedure

There are several software packages available for general-purpose MBD simulations. SIMPACK [3] is used for the example vehicle simulation given here.

TABLE 3.1
Examples of commercially available vehicle motion analysis programmes (as of 2024, according to the author)

Name	Manufacturer
SIMPACK	Dassault Systemes (France)
VI-RAIL	VI-grade (Germany)
VAMPIRE-PRO	Resonate (UK)
NUCARS	MxRAIL (USA)
VOCO	IFSSTAR (France)
GENSYS	DESolver (Sweden)

Note: All company and product names mentioned in the table are trademarks or trade names of the companies concerned.

Figure 3.2 shows the flow from building the vehicle model using SIMPACK to carrying out the curving simulation.

To build a vehicle model, the rail cross-sectional profile, tie-plate angle, gauge, rail support stiffness and other properties are set. Figure 3.3 illustrates typical wheel tread and rails in Japan. Various rail cross-sectional shapes, including worn profiles, can be handled. Next, a wheelset is defined on the rail. The parameters of the wheelset include mass, moment of inertia, tread profile, back-gauge and wheel radius in the neutral position. The coefficient of friction between the wheel and rail is also set at this time. There are also a number of models available to represent the creep force between wheel and rail, but the creep force calculation by the programme FASTSIM proposed by Kalker [4,5] is generally used.

A bogie model is then created by connecting the bogie frame to the axlebox position of the wheelset using the axle support mechanism. Parameters such as spring constants are selected, and the mass, moment of inertia and centre of gravity of the bogie frame are set. If the bogie is a two-axle bogie, the model can be run once to check the operation by users.

The car body is defined as a rigid element and its mass; moment of inertia and centre of gravity are set. A vehicle model is completed by adding support elements such as air springs, traction links, side dampers and stoppers that connect the vehicle body to the bogie frame. For bogies with bolsters, the bolsters and side beams should be modelled with care. Recently, the car body can also be included in the model as an elastic element, but this is time consuming during simulation execution and should be applied only when the problem being considered requires it.

Track parameters such as curve length, curve radius, transition curve shape, cant, type of curve with cant, and slack are entered for the vehicle running simulation. The support stiffness of the track can also be selected. In some cases, track irregularities are applied in the simulation. The vehicle runs with a constant

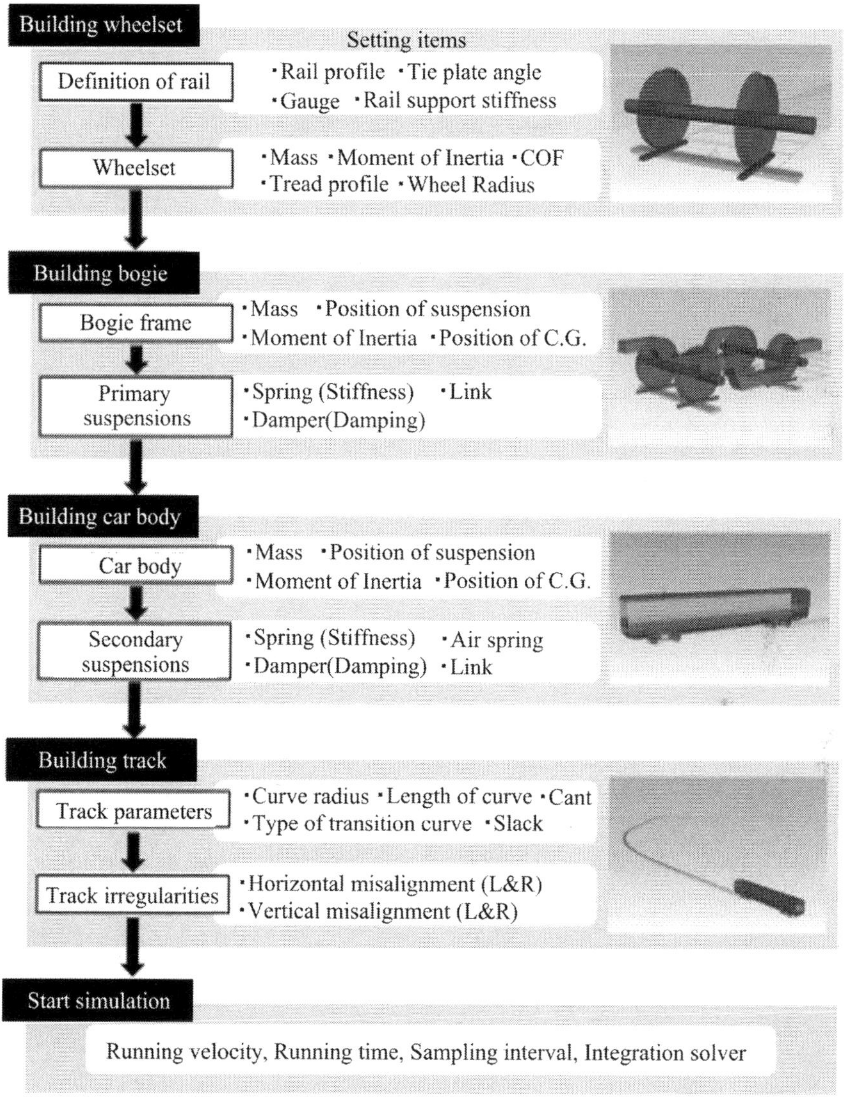

FIGURE 3.2 Flow of vehicle modelling (SIMPACK example).

speed or a predetermined speed pattern on the input track. In this case, the front and rear DOFs of the car body do not exist. The vehicle can also simulate actual running conditions by applying a prescribed torque to the wheelset. Numerical integration is performed by entering the above conditions. Numerical integration methods with variable time steps are common, and the recommended numerical integration method is often set by default for each software package.

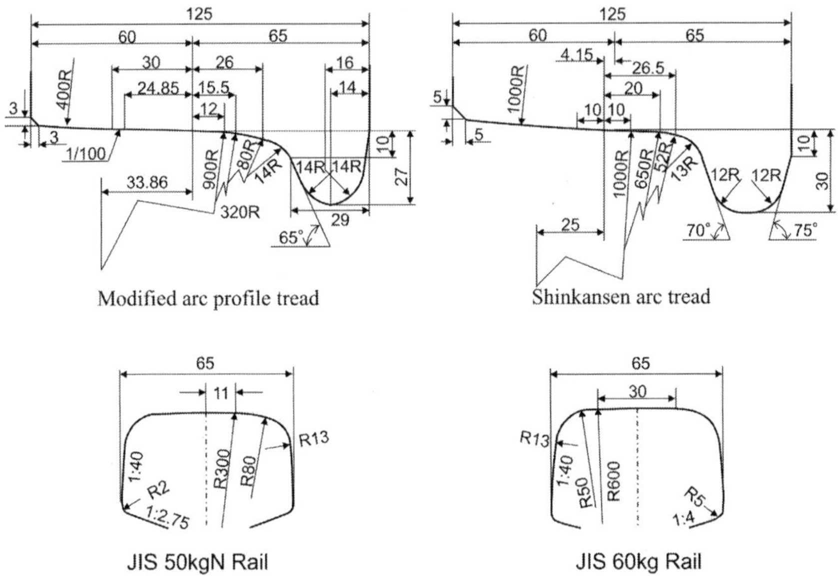

FIGURE 3.3 Typical Japanese wheel treads and rail head profiles.

In running simulations, the time response is calculated using the numerical integration method, and it is convenient to take the distance on the horizontal axis into account, given the vehicle velocity. This facilitates analysis of ride comfort, such as where on the track the vehicle vibration increases, and discussion of running safety, such as where the wheel load and lateral force change. The results of the analysis can also be shown graphically. The user interface is well-developed and allows the user to visually check the vehicle movement and the wheel–rail contact position.

3.2.2 Curving Simulation

An example of curving simulation is shown in Figure 3.4, where the lateral force, wheel load and derailment coefficient are calculated when the vehicle runs on a curved track whose curve radius is R = 200 m. For the wheel tread geometry and rail geometry, the combination of modified arc tread and JIS 50 kgN rail was chosen. One vehicle is modelled with seven rigid bodies, each with 6 DOFs, for a total of 42 DOFs, under constant running speed conditions of 25 km/h. This type of running simulation allows the running safety of the designed vehicle to be quantitatively investigated and can be used to optimize various parameters to further improve performance. In addition, graphical confirmation of the wheel-rail contact positions is helpful in gaining a deeper understanding of the phenomenon.

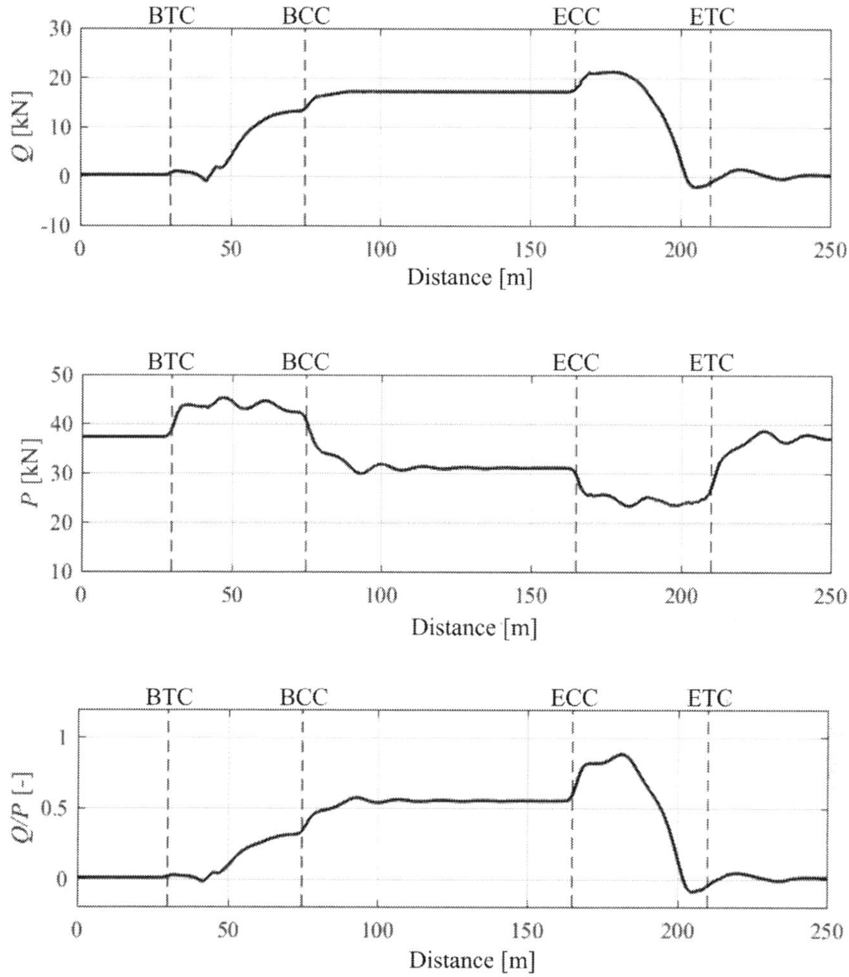

FIGURE 3.4 Example of simulation results.

Figure 3.5 is an example of the illustration of the contact points among four wheels in a bogie as it runs on a sharp curve. As shown in the figure, the wheel–rail contact points are different among the four wheels in a bogie during curving. Generally, the outside wheel of the leading wheelset contacts the rail at the flange, while both wheels of the trailing wheelset contact the tread near the neutral position. Especially in the curving simulation, the contact condition of the leading outside wheel not only has lateral and vertical forces, but also the flange contact angle shown in Figure 3.6 plays an important role when evaluating running safety against flange climb derailment.

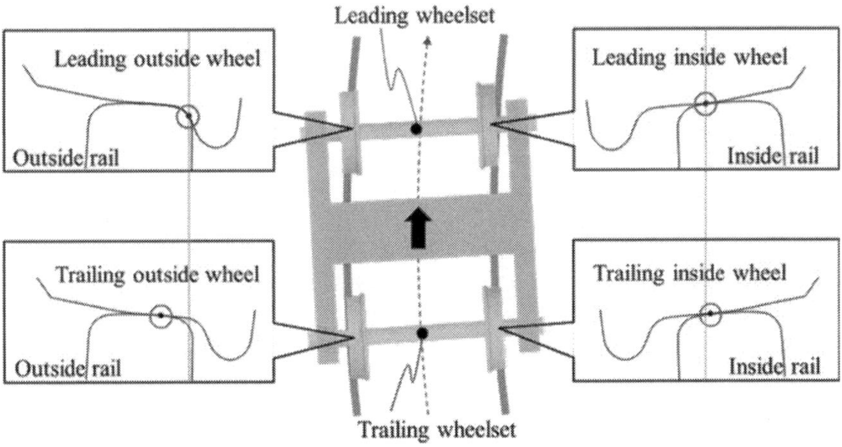

FIGURE 3.5 Contact points when the vehicle runs on a sharp curve.

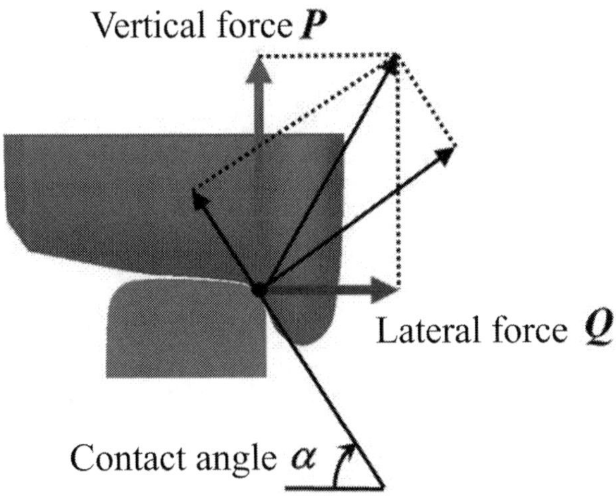

FIGURE 3.6 Contact forces and contact angle between the wheel and rail.

3.2.3 Safety Assessment of Flange Climb Derailment

Flange climb derailments sometimes result in serious accidents, and various studies for their prevention have been carried out around the world. Elkins and Cater described the safety assessment methods from European, British and North American viewpoints [6]. In European countries, safety estimation has been studied, such as in a Netherlands International Union of Railways ORE report [7], and the guidelines for the prevention of flange climb derailments are provided in European Standard EN14363 [8]. In the United States, TTCI engineers described the limit criterion against flange climb derailments in a handbook [9]. In these guidelines, the evaluation and the limit value for safety against derailment are provided as a *Y*/*Q* or *L*/*V* value, which is the ratio of the lateral force *Y* or *L* to the vertical force *Q* or *V*, and given by Nadal's formula as the function of flange angle α and coefficient of friction μ:

$$\left(\frac{Y}{Q}\right)=\frac{\tan\alpha-\mu}{1+\mu\tan\alpha}. \tag{3.1}$$

In Japan, the safety evaluation and the limit value against derailment is the *Q*/*P* value, which is equivalent to the *Y*/*Q* or *L*/*V* value [10, 11]. Table 3.2 shows these notations in different countries and regions.

The safety measure stipulated by the Japan Ministry of Land, Infrastructure, Transport and Tourism (MLIT) is the 'Estimated *Q*/*P* ratio', which is defined as the ratio of the critical value to estimated value of derailment coefficients *Q*/*P* of the leading outside wheel of the bogie:

$$\text{'Estimated } Q/P \text{ ratio'} = (Q/P)_{cri}/(Q/P)_{est}. \tag{3.2}$$

where $(Q/P)_{est}$ is the estimated derailment coefficient calculated by the ratio of the estimated lateral contact force Q_{est} to the estimated vertical contact force P_{est}. In this process, Q_{est} is calculated considering parameters such as the friction coefficient, unbalanced centrifugal force and yaw resistance of the bogie; P_{est} is calculated considering the unbalanced centrifugal force and track twist, and the critical $(Q/P)_{cri}$ is calculated by a modified Nadal's equation as follows:

$$\left(\frac{Q}{P}\right)_{cri}=\frac{\tan\alpha-\mu_e}{1+\mu_e\tan\alpha}. \tag{3.3}$$

using the equivalent friction coefficient μ_e, which increases with increasing angle of attack. In the MLIT's procedure [12], the angle of attack value is given by a chart created from the results of numerical simulations as a function of track

TABLE 3.2
Derailment prevention parameters used by various countries

Country or Region	Lateral force	Vertical force	Derailment coefficient
Japan	*Q*	*P*	*Q*/*P*
Europe and North America	*Y*	*Q*	*Y*/*Q*
UK	*L*	*V*	*L*/*V*

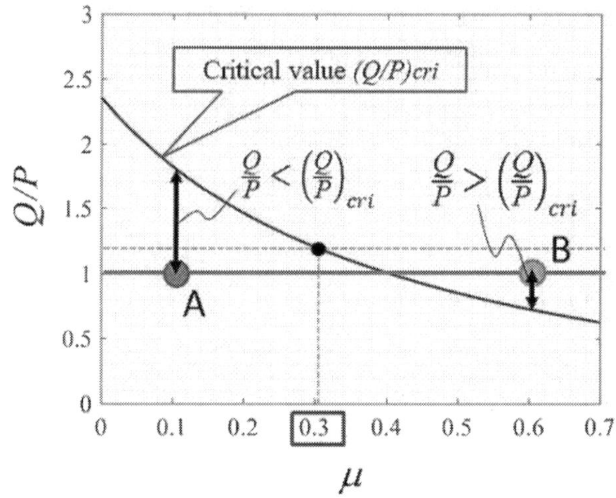

FIGURE 3.7 Relationship between the real Q/P value and the critical value of Q/P (flange angle = 67°).

curvature (1 / m). The philosophy of this safety measure is effective and can be used to prevent derailments, but it has a number of issues related to the assumed value of the coefficient of friction. For example, the friction coefficient μ between the flange surface and the rail head changes when the lubrication and other conditions change.

Figure 3.7 shows the relationship of the measured *Q/P* against the critical value of *Q/P*. In this figure, the critical values of *Q/P* are drawn in the relationship to the friction coefficient between flange and rail. At point B in the figure, the value of *Q/P* is greater than the critical value, and it shows that the risk of flange climb derailment may be increased. Contrarily, at point A in the figure, the value of *Q/P* is smaller than the critical value and it is considered to indicate safety against flange climb derailment. In this way, the value of *Q/P* cannot serve as a safety estimation only by itself, and the ratio to the critical value is more important. Therefore, the Flange Climb Index (FCI) was introduced as a measure of safety against flange climb derailments [13].

$$\mathrm{FCI} = \frac{(Q/P)}{(Q/P)_{\mathrm{cri}}}. \tag{3.4}$$

In cases where the FCI is larger than 1.0, flange climb starts as shown in Figure 3.8. The figure shows the calculated FCI for a curve with a radius of 160 m for $\mu = 0.35$ and $\mu = 0.65$ by MBD simulation. For $\mu = 0.65$, the FCI increases immediately after the train enters the exit-side transition curve and exceeds 1.0. Subsequently,

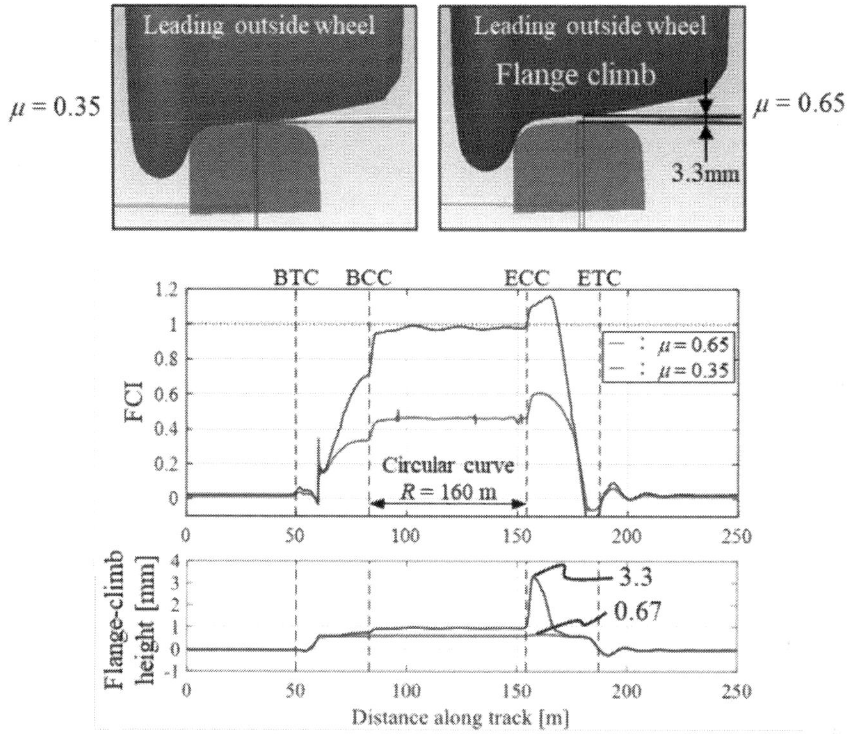

FIGURE 3.8 Flange climb height and flange climb index.

a flange climb of 3.3 mm occurs near the instance of maximum FCI. In contrast, flange climb does not occur in the case of $\mu = 0.35$, where the FCI value does not exceed 0.6. Thus, the FCI is effective for determining the possibility of flange climb. To calculate the FCI value, it is necessary to determine the actual friction coefficient value between the wheel flange and rail. A way to estimate the value using the PQ monitoring bogie [14], which measures the wheel load P and lateral force Q, is described in Chapter 4, but it still remains a difficult problem.

3.2.4 Example of Wheel Climb Derailment Accident

Tokyo Metro experienced a derailment accident on the Hibiya line in March 2000. After the derailment accident, several experiments were conducted to investigate its cause. With great difficulty, it was determined that the cause of the derailment was a multi-factor derailment caused by the track, the vehicle and the interaction between the track and the vehicle.

One of the causes of the derailment was thought to be the increased coefficient of friction between the wheel and the rail on sharp curves with a large angle of attack, resulting in a large lateral force. Figure 3.9 shows the results of measuring

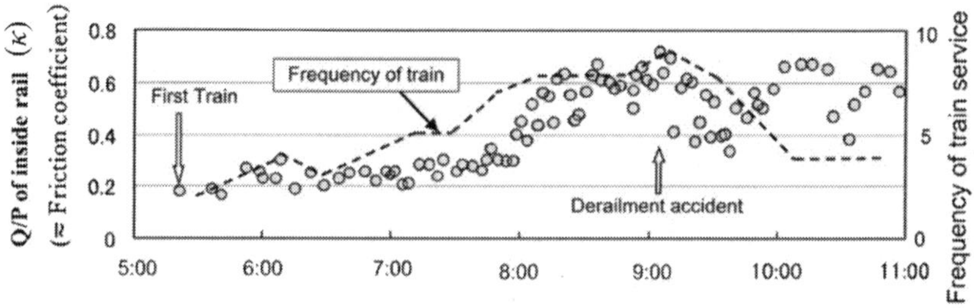

FIGURE 3.9 Change in the inside Q/P on the sharp curve of the Hibiya line accident site.

the ratio of the lateral force (Q) to the vertical force (P) for the inner side wheel, which corresponds to the coefficient of friction from 5:00 to 11:00 for running trains. It is estimated that the coefficient of friction was around 0.2 when the first train of the day ran on the track, but then increased, reaching a maximum of around 0.7 at 9:00, when the derailment occurred. This increase in the friction coefficient was thought to have led to an increase in the lateral force and an increase in the derailment coefficient. Surprisingly, the coefficient of friction between the wheel and the rail and the associated derailment coefficient were found to vary from time to time. These causes of derailments led to two countermeasures. One was to develop technology to monitor the friction coefficient and the derailment coefficient on the track at all times. The other was to reduce the angle of attack to reduce lateral forces through the development of the single-axle steering bogie.

3.3 DYNAMICS OF TRAIN OVERTURN

When running on curves at velocities far in excess of the speed limit or when crosswinds are strong, rail vehicles are subjected to large forces that can cause overturn derailment. In this section, an evaluation index and some case studies are described.

3.3.1 Basic Mechanism

Figure 3.10 shows the process of overturn derailment of a railway vehicle. When rail vehicles are subjected to the large forces described above, first, one wheel can come off the rail and then, as that wheel rises, the flange of the opposite wheel rides up onto the rail and derails off the track just before it falls over. Such derailments are specifically referred to as overturns because the causes and mechanisms are different from those of flange climb derailments, and their safety is assessed in a different way. Specifically, the rate of off-loading is used as an evaluation index, and the overturn limit is set when the rate of wheel load reduction on one side is 1.0, that is, when the wheel load is zero. When the wheel load on the upwind side

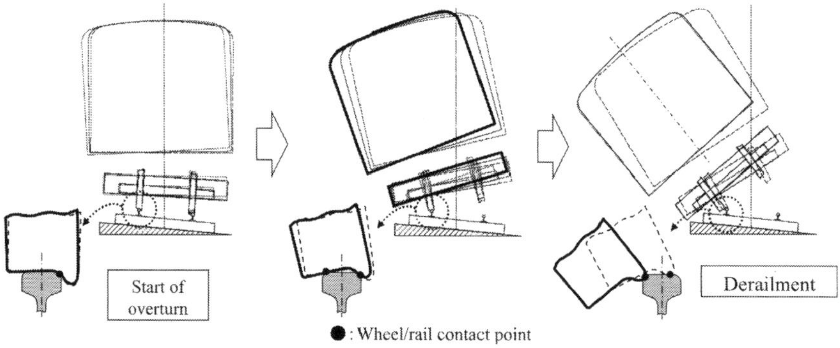

FIGURE 3.10 Train overturn sequence.

of a vehicle in a crosswind becomes zero, the vehicle is considered to be at the limit of stability against an overturn, and the wind speed at this point is specifically referred to as the critical wind speed for overturning.

3.3.2 Kunieda's Formula

Kunieda's formula [15] is a simplified formula to calculate the risk of a train overturning, considering the centrifugal force of running on a curve, the vehicle and lateral vibration and the crosswind effect. It is a simplified equation based on a static mechanism, but it agrees well with real phenomena. It has been certificated in many cases and used for safety guidelines in Japan.

Figure 3.11 shows an overview of the dynamic model of train overturn. Danger index D against overturning is calculated by the following equation:

$$D = \frac{2}{G}\left\{h_G^* \frac{a_u}{g} + h_G^*\left(1 - \frac{\mu}{1+\mu}\frac{h_{GT}}{h_G^*}\right)\frac{a_y}{g} + \frac{h_{BC}^* F_S}{(1+\mu)m_B g}\right\}. \tag{3.5}$$

The first term of the formula describes the effect of the gravity and the centrifugal force considering the cant (superelevation), the second term shows the lateral vehicle vibration considering the mass ratio of the body and bogies and the third term shows the effect of crosswind. The third term of the equation expresses aerodynamic forces, and a detailed explanation is omitted here. Equation (3.5) can be simplified as follows:

$$D = \frac{2}{G}\left\{h_G^* \frac{a_u}{g} + h_G^*\left(1 - \frac{\mu}{1+\mu}\frac{h_{GT}}{h_G^*}\right)\frac{a_y}{g}\right\}. \tag{3.6}$$

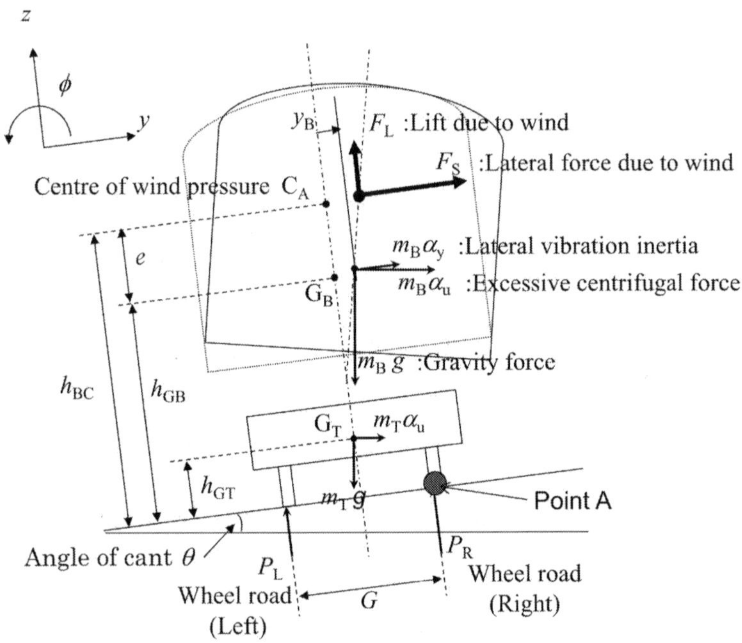

FIGURE 3.11 Definitions of train overturn parameters.

where h_G^* (= $1.25h_G$) is the compensated value of the nominal height of the gravity centre h_G of the vehicle considering the distortion of bogie springs shown in Figure 3.11, μ is the mass ratio of two bogies/(body + passengers), h_{GT} is the C.G height of a bogie, α_y is the acceleration of vehicle lateral vibration ($\alpha_y = 0.1g, v \geq 80$ km/h, α_y=$0.1g \times v/80, v \leq 80$ km/h) and a_u is the excessive centrifugal force defined as follows:

$$a_u = g\left(\frac{a_r}{g} - \frac{C}{G}\right) = g\left(\frac{v^2}{Rg} - \frac{C}{G}\right). \tag{3.7}$$

The critical speed v against overturn is calculated by setting D as 1 in Equation (3.6).

3.3.3 Example of Numerical Simulation

To compare the numerically obtained velocity that shows a train overturning, the results by simulations are compared with the values obtained by Kunieda's formula. In the numerical simulation, the value of the wheel load is evaluated to

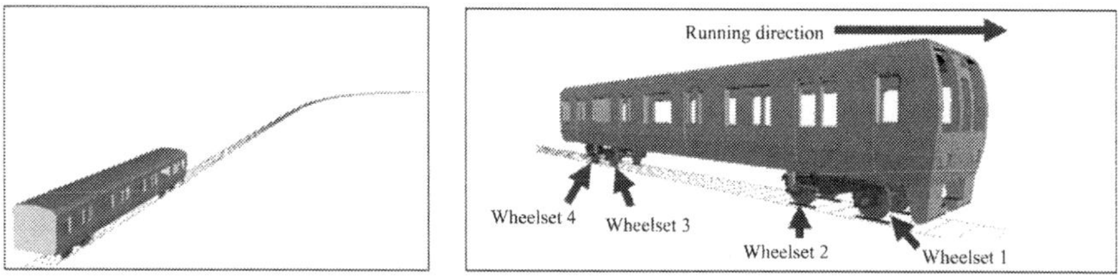

FIGURE 3.12 Overturning simulation model.

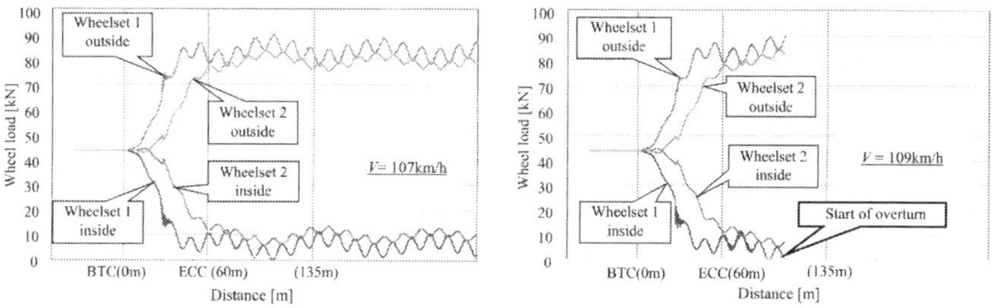

FIGURE 3.13 Overturning simulation results.

detect the overturning vehicle velocity. When a wheel indicates zero wheel load, the train is considered as having reached the overturn velocity.

Figure 3.12 shows an example of numerical simulation when a vehicle runs at 80 km/h on a track without irregularities. The vertical axis shows the load for all wheels of the vehicle. As shown in the simulation, the load for the leading inside wheel is the smallest due to the ability of the vehicle to travel on a curve. At this velocity, it retains enough wheel load against overturn derailment. Figure 3.13 shows an example when the vehicle runs at 109 km/h without track irregularities. The load for the leading inside wheel is close to zero, which means that the vehicle velocity is at the critical limit for overturn derailment.

To evaluate the fundamental effectiveness of Kunieda's formula for different gauges, some adjusted numerical simulation models are used for verification. In the numerical simulation, the gauges are adjusted as 1.435 and 1.667 m to consider both standard and wide gauges. The lateral distance of the attachment points of primary suspensions are suitably adjusted in these brief simulation conditions.

Table 3.3 shows the results of a comparison between Kunieda's formula and MBD simulation in terms of the critical velocity limit against train overturn. The agreement between them is good so we can conclude that Kunieda's formula is fundamentally useful for predicting the actual train overturn speed almost corresponding to the MBD simulation.

TABLE 3.3
Comparison of Kunieda's formula and MBD simulation for critical speed limit against overturning

Track gauge	Kunieda's formula	MBD simulation
Narrow gauge (1.067 m)	105.7 km/h	109 km/h
Standard gauge (1.435 m)	118.7 km/h	121 km/h
Wide gauge (1.667 m)	127.0 km/h	130 km/h

3.3.4 Examples of Train Overturn Accident Analysis [16]

Derailment accidents due to excessive speed on curves have occurred around the world, and the results of analysis of some of these accidents are discussed here. Table 3.4 shows features such as the track gauge, curving radius, train running speed and speed limit at accident sites. The estimated limit speed against overturn is calculated by the investigation organizations and/or by one of the authors, using numerical simulations and/or the simplified formula mentioned before.

In the case of the Swiss Glacier Express, the overturn was caused by too early acceleration at the exit of a curved section, so the trailing vehicles in the train set exceeded the critical speed against train overturn. This shows that it is also important to consider the running speed at the exit of curves, especially for a long train set and high-acceleration vehicles, including the case of passing through turnouts.

At all accident sites, no speed limit signalling devices were provided. The running speeds at the accident sites were +21 to +83 higher than the regulated speed limits and +3 to +30 higher than the critical speeds.

Parameters such as the track gauge width, the CG (Centre of Gravity) height of the vehicle and the curving radius mainly determine the critical speed limit against overturn. The relationship between the critical speed and these parameters can be calculated by Kunieda's formula. The critical speed is heavily influenced by the ratio of the CG height of the vehicle to the track gauge width, which is called the CG/gauge ratio. The relationship between the critical speed limit against overturn and CG/gauge ratio is depicted in Figure 3.14. The features of the train overturn accidents in Table 3.4 are also plotted in the figure. The overturn critical speed limit V_{cri} decreases as the CG height/gauge width ratio increases, that is, due to an increase in the CG height of the vehicle, and the track gauge width decreases. A decrease in the curve radius R also decreases the critical speed.

Figure 3.15 illustrates an overview of the critical speed limit to prevent train overturn against the CG height/gauge ratio and the curving radius in a three-dimensional graph. The dot in the centre of the graph shows a condition almost equivalent to that at the Fukuchiyama accident site. The critical speed increases as the cant angle increases, which is the ratio of super-elevation versus the gauge

TABLE 3.4
Outline of train overturn accidents

Accident name (Country)	Kind of train	Gauge (mm)	Curve radius (m)	Speed (km/h)	Speed limit (km/h)	Estimated limit speed against overturn (km/h)
Fukuchiyama L. (Japan)	Commuter (EMUs)	1067	304	116	70	106–108
Glacier Express (Swiss)	Sight-seeing (EL+PCs)	1000	85	56	35	52–55
Spanish high-speed train (Spain)	High-speed (composed train set)	1668	380	153	80	130–145*
NY Metro North Rail (USA)	Commuter (push-pull)	1435	291	131	48	115–125*
Amtrak Philadelphia (USA)	High speed (EL+PCs)	1435	437	163	80	130–140*

* Rough estimate.

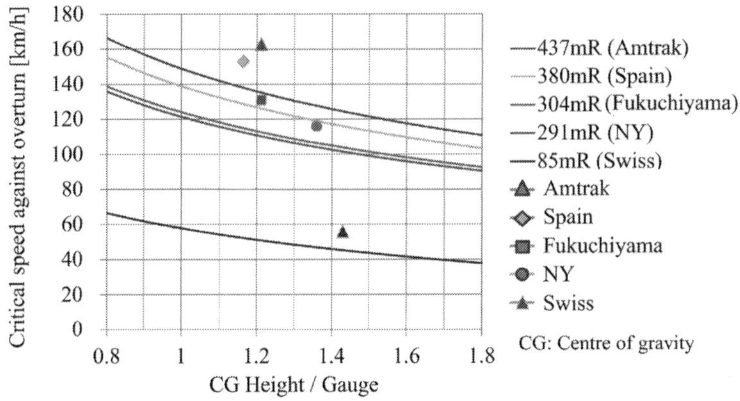

FIGURE 3.14 Comparison of train overturn accidents.

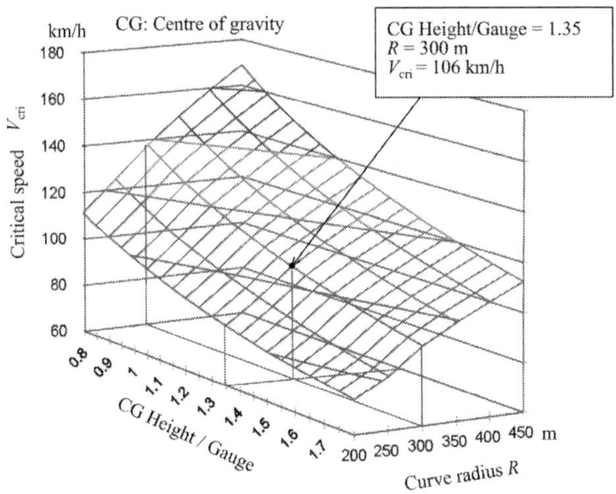

FIGURE 3.15 Overview of train overturn critical speed limit.

width. The number of passengers also influences the CG height, so the increase in the number of passengers causes the critical speed limit to decrease.

The influences of these factors are shown in Figure 3.16 around the reference point (V_{cri} = 106 km/h; C.G/gauge = 1.35, R = 300 m, c = 0.1 m and 92 passengers). As for track super-elevation, a decrease of c = 0.1 m to 0 m decreases the critical speed V_{cri} = 106 km/h to 88 km/h, and an increase of c = 0.1 m to 0.15 m increases V_{cri} = 106 km/h to 114 km/h. As for passenger numbers, an increase to 150% of full capacity (from 92 to 210 passengers) decreases the critical speed V_{cri} = 106 km/h to 103 km/h, and a decrease to no passengers (from 92 to 0) increases V_{cri} = 106 km/h to 109 km/h.

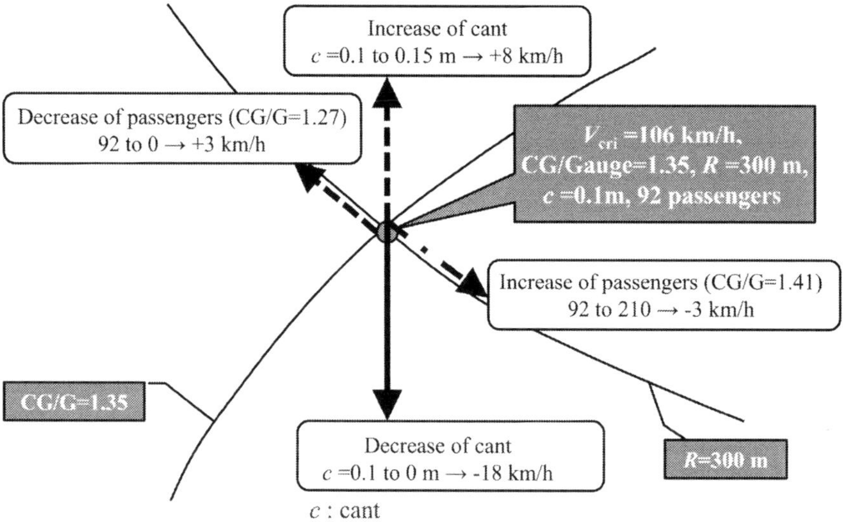

FIGURE 3.16 Factors influencing train overturn critical speed.

Although the effects of wind are not taken into account in this book, various studies have been conducted on safety assessments that take wind into account [17–20].

REFERENCES

1. A.A. Shabana, K.E. Zaazaa, H. Sugiyama, *Railroad Vehicle Dynamics A Computational Approach*, CRC Press, 2007.
2. S. Iwnicki, *The Manchester Benchmarks for Rail Vehicle Simulation*, Taylor & Francis Group, 1999.
3. G. Schupp, H. Netter, L. Mauer, M. Gretzschell, *Multibody System Simulation of Railway Vehicles with Simpack*, The Manchester Benchmarks for Rail Vehicle Simulation, 1999.
4. J.J. Kalker, A fast algorithm for simplified theory of rolling contact, *Vehicle System Dynamics*, 11, 1–13, 1982.
5. J.J. Kalker, *Three-Dimensional Elastic Bodies in Rolling Contact, Solid Mechanics and its Applications*, Vol. 2, Kluwer Academic, 1990.
6. J.A. Elkins, A. Carter, Testing and analysis techniques for safety assessment of rail vehicles: The state-of-the-art, *Vehicle System Dynamics*, 22(3–4), 185–208, 1993.
7. ORE, Permissible maximum values for the Y- and Q- forces and derailment criteria, Question C138, *Report C138, Report No.9 (Final report)*, 1986.
8. CEN (European Committee for Standardization), Railway applications: Testing for the acceptance of running characteristics of railway vehicles; Testing of running behaviour and stationary tests, *European standard EN 14363*, 2005.
9. W. Huimin, W. Nicholas, Railway vehicle derailment and prevention, *Handbook of Railway Vehicle Dynamics, Chapter 8*, CRC Press, Taylor & Francis, 209–237, 2006.

10. N. Wilson, R. Fries, M. Witte, A. Haigermoser, M. Wrang, J. Evans, A. Orlova, Assessment of safety against derailment using simulations and vehicle acceptance tests: A worldwide comparison of state-of-the-art assessment methods, *Vehicle System Dynamics*, 49(7), 1113–1157, 2011.
11. H. Ishida, T. Miyamoto, E. Maebashi, H. Doi, K. Iida, A. Furukawa, Safety assessment for flange climb derailment of trains running at low speeds on sharp curves, *Quarterly Report of RTRI*, 47(2), 65–71, 2006.
12. H. Takai, M. Uchida, H. Muramatsu, H. Ishida, Derailment safety evaluation by analytic equations, *Quarterly Report of RTRI*, 43(3), 119–124, 2002.
13. A. Matsumoto, Y. Michitsuji, Y. Ichiyanagi, Y. Sato, H. Ohno, M. Tanimoto, A. Iwamoto, T. Nakai, Safety measures against flange-climb derailment in sharp curve considering friction coefficient between wheel and rail, *Wear*, 15, 432–433, 2019.
14. A. Matusmoto, M. Tomeoka, K. Matsumoto, M. Tanimoto, T. Nakai, A new measuring method of wheel-rail contact forces and related considerations, *Wear* 265, 1518–1525, 2008.
15. M. Kunieda, Theoretical study on the mechanics of overturn of railway rolling stock, *Railway Technical Research Report, No.793*, 1972 (in Japanese).
16. A. Matsumoto, Y. Michitsuji, Y. Tobita, Analysis of train-overturn derailments caused by excessive curving speed, *International Journal of Railway Technology*, 5(2), 27–45, 2016.
17. G. Matschke, P. Deeg, B. Schulte-Wering, Effects of strong cross winds on high-speed trains, a methodology for risk assessment and development of countermeasures, *Proceedings of WCRR 2001*, Fraunhofer, 2001.
18. P.E. Gautier, Strong wind risks in railways: the DEUFRAKO crosswind program, *Proceedings of WCRR 2003*, Fraunhofer, 463–474, 2003.
19. Y. Hibino, H. Ishida, Static analysis on railway vehicle overturning under crosswind, *RTRI Report*, 17(4), 39–44, 2003 (in Japanese).
20. Y. Hibino, T. Shimomura, K. Tanifuji, Verification of static analysis on railway vehicle overturning under crosswind, *Transaction of Japan Society of Mechanical Engineers, Series C*, 75(758), 2605–2612, 2009 (in Japanese).

4 Condition Monitoring and Condition-Based Maintenance

4.1 GENERAL CONCEPTS AND METHODS OF CONDITION MONITORING

Condition monitoring is considered to be an advanced concept for Fault Detection and Isolation or Identification (FDI), which has been the subject of various studies [1]. Condition monitoring is mainly applicable to systems whose condition degrades over time. The main purpose of condition monitoring is to detect faults before they occur and to identify their causes [2, 3].

Figure 4.1 illustrates the concept of condition monitoring. Diagnosis can be performed on time-series data obtained from various sensors by focusing on the current information. In addition, the future state can be predicted (to formulate a prognosis) by analysing the changes in data from the past.

The general framework of state monitoring or FDI is shown in Figure 4.2. The outputs are the measurements and the inputs correspond to those measurements from an asset. Unobservable disturbances can affect the system. In general, we consider a dynamic system failure corresponding to an input $\boldsymbol{u}$, output z and unknown disturbance $\boldsymbol{d}$.

4.2 MODEL-BASED METHODS

If a physical model of the asset and the input $\boldsymbol{u}$ are available, the model-based method, which can be defined as detection and decision-making based on the evaluation of residuals, can be used. Figure 4.3 shows the basic structure of a model-based FDI system.

Residuals can be generated in a number of ways, and they essentially capture the differences between what the model and the measurements indicate. They should be zero or close to zero if the model and the system itself are similar, that is, if there is no fault. Evaluation of the deviations away from zero can then be used to determine fault(s).

DOI: 10.1201/9781003362135-4

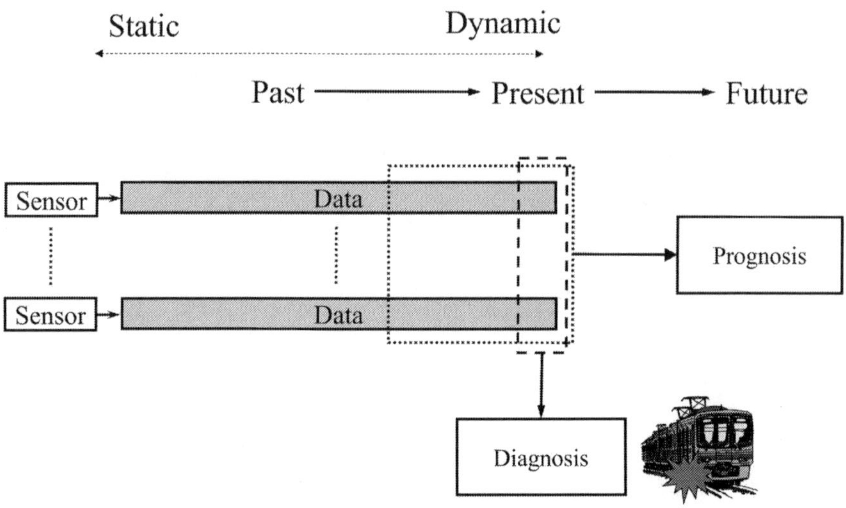

FIGURE 4.1 Concept of condition monitoring.

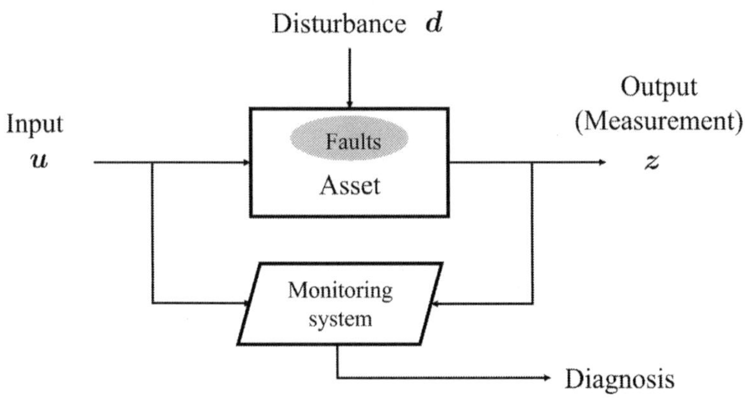

FIGURE 4.2 General framework of FDI.

Model-based FDI can be divided into the following methods:

- Parameter estimation method
- Parity equation method
- Observer/Kalman Filter (KF)-based method.

4.2.1 Parameter Estimation Method

Parameter estimation is applicable if the faults are associated with changes in the system parameters. The residuals are the differences between the nominal system model parameters and the estimated parameters.

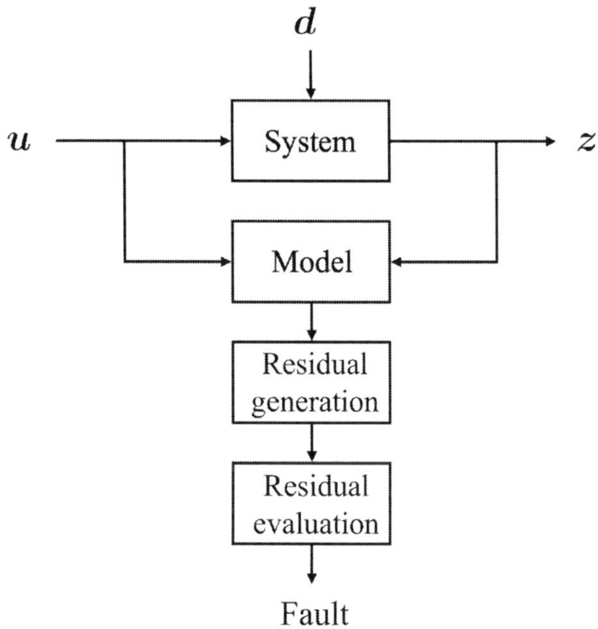

FIGURE 4.3 Basic structure of model-based FDI system.

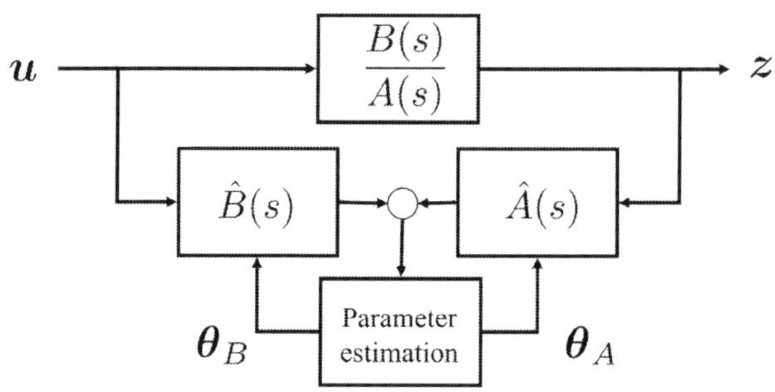

FIGURE 4.4 Parameter estimation with the equation error method.

Various parameter estimation methods are possible, such as the equation error method and output error method, which are shown schematically in Figures 4.4 and 4.5, respectively.

Fault detection by parameter estimation relies on the fact that a possible fault in the monitored system can be associated with specific parameters in the modelled system given by an input and output relation.

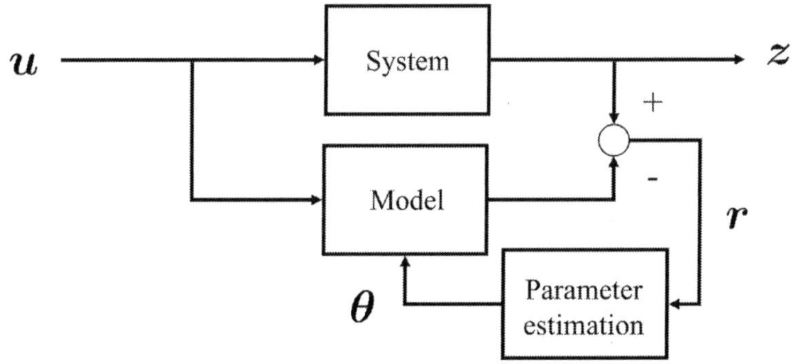

FIGURE 4.5 Parameter estimation with the output error method.

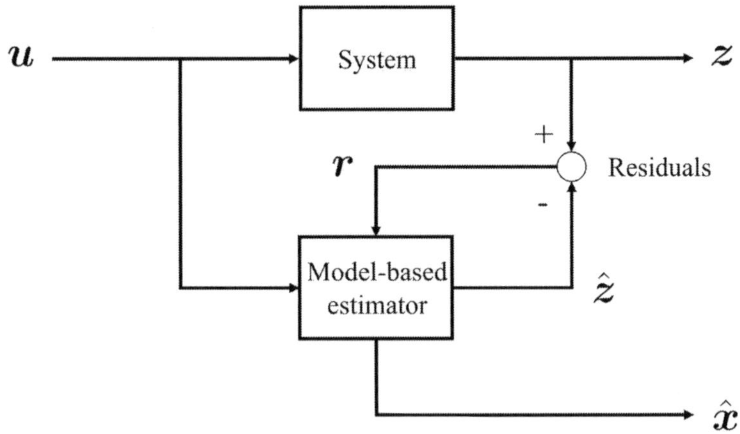

FIGURE 4.6 Model-based state estimator.

4.2.2 Observer/Kalman Filter-Based Method

The existence of a physical model enables an observer-based method to be used. The residuals, $\boldsymbol{r}$, are obtained by the difference between the measured output, z, and the predicted output from the nominal model, as shown in Figure 4.6.

These methods include observers, KFs (see Appendix A2.6), Extended KFs (EKFs) (see Appendix A2.7), unscented KFs and particle filters. Cubature KFs have also received attention in recent years.

The observer-based method is appropriate if the faults are associated with unmeasured state variables. The unmeasured states are constructed from the measurable input and output using a Luenberger observer for the deterministic case or a KF for the stochastic case.

The particle filter is a method that can handle nonlinear systems, but it has the disadvantage of a large computational load. KFs are used in many fields due to their low computational load and suitability for real-time data processing.

4.3 MODEL-FREE METHODS

If the only output available is z, signal processing methods without models can be applied. These methods include frequency selective filters, spectral analysis, the maximum entropy method, wavelet analysis and vibration analysis (on the amplitude and phase). Other methods include expert systems, fault tree analysis, principal component analysis and independent component analysis.

Recently, techniques for analysing large amounts of data using artificial intelligence are spreading. In addition to traditional machine learning techniques such as neural networks and support vector machines, deep learning using deep neural networks, which can perform unsupervised learning, is rapidly developing. These techniques are also used in current railway condition monitoring, and their use is expected to grow in the future.

4.4 APPLICATION TO RAILWAY CONDITION MONITORING

4.4.1 Concept of Railway Condition Monitoring and Literature Review

The purpose of condition monitoring in railways is to monitor the condition of vehicle components and track and detect signs of impending derailment. Figure 4.7 shows the relationship between inputs and outputs for condition monitoring on railways (mainly for rolling stock and track systems). The track geometry is input and the vehicle vibration, noise and temperature are output. The objective is to use this information to detect the condition of the vehicle suspension, signs of impending derailment and track conditions. In particular, track geometry cannot be measured directly, so it must be treated as an unknown state to be estimated.

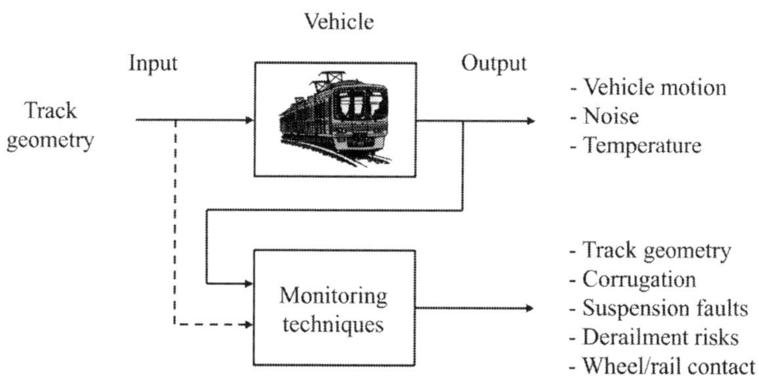

FIGURE 4.7 Condition monitoring in railway.

When constructing a diagnostic system, particular attention should be paid to the following questions:

- What do we want to diagnose and when?
- What should be measured?
- What can be measured?
- Is sensor fusion necessary?
- Can we identify changes in conditions over time?
- What is the relationship between degradation trends and the measured data?
- How do we extract hidden signals?
- How do we process huge amounts of data?

4.4.2 Vehicle Condition Monitoring

4.4.2.1 Detection of Defects and Deterioration in Vehicle Suspensions

Vehicle inspections are particularly important to prevent serious accidents. Condition monitoring is necessary for early detection of vehicle failures, and methods to detect vehicle defects based on signals from sensors attached to the vehicle have been proposed for real-time monitoring. Condition monitoring is an important element of Condition-Based Maintenance (CBM), which detects and identifies deterioration before it causes a failure. FDI technology, which has been studied extensively, can be applied to vehicle condition monitoring [4]. Li et al. provided an overview of approaches applied to railway vehicle on-board health monitoring systems [5]. Charles et al. described a method to estimate wheel/rail profile and adhesion state using a KF [6], and a Rao-Blackwellised particle filter to estimate damping coefficients and tread gradients of left-right dynamic and yaw dampers [7]. The EKF has been applied to health monitoring of the lateral suspension of the Italian ETR 500 class of high-speed trains [8]. A model-based condition monitoring strategy based on the recursive least-square algorithm was proposed by Liu et al. [9].

Jesussek et al. described a method to detect left and right suspension anomalies and their types using a hybrid EKF [10]. Tsunashima et al. developed a method to detect suspension failures using the interacting multiple model method [11, 12]. Moreover, a method for detecting suspension failures using multiple models without considering dynamic transitions between models has also been reported [13]. A method for monitoring the condition of railway vehicle suspensions using the cubature KF has also been proposed [14].

For monitoring the condition of suspension systems, an online modal identification scheme was proposed by Liu et al. [15]. Health monitoring of high-speed train suspension using computational dynamics and acceleration measurements was described by Lebel et al. [16]. A robust data-driven method for on-board vibration-based degradation detection in railway suspensions was introduced in [17]. A multistage clustering framework for unsupervised condition monitoring of train wheels was proposed by Ghiasi et al. [18]

4.4.2.2 Detection of Improper Running Conditions

Vehicle condition monitoring can also be used to detect inappropriate running conditions such as derailments, hunting and temperature increases in the axlebox that may affect running safety. In particular, derailments require early detection of signs of derailment in order to take appropriate action before the situation reaches an unrecoverable stage. Sun et al. describe a measuring system for hunting behaviour monitoring of high-speed railway vehicles using signal-based techniques [19].

Wheel load and lateral forces, which are physical quantities that express the condition between the wheels and rails when a train is running, are very important control items for the safety management of railways. The lateral force divided by the wheel load is called the derailment coefficient, and this value is controlled to be less than a standard value to prevent derailments [20]. Wheel load and lateral force have conventionally been measured by affixing strain gauges to the wheels and extracting them using slip rings, or by using strain gauges and displacement gauges on the rails on the ground side.

Xia et al. [21] proposed a method to estimate derailment coefficients using an inverse model, and Matsumoto et al. [22] developed a non-contact method to estimate derailment coefficients using an optical sensor continuously mounted on the vehicle. Tokyo Metro has developed a monitoring bogie that can measure the derailment coefficient in a non-contact manner as described above, and report daily and monthly changes in the derailment coefficient [23, 24].

However, it is also important to take action to deal with the post-derailment consequences. This is not aimed at preventing derailments, but it is important for reducing secondary risks, particularly for freight trains. Boronenko et al. [25] described a method for detecting derailments of freight vehicles from bogie vibrations, which was validated by numerical simulation and in-service train testing. Hubacher et al. [26] proposed a method for emergency braking to act on freight trains. Koga et al. reported on the development of a system for detecting improper running of vehicles [27].

The coefficient of friction between the wheel and rail is an important factor in the longitudinal control of a vehicle. Ward et al. [28] and Hubbard et al. [29] developed a method for on-board detection of low adhesion conditions using a Kalman–Bucy filter and a signal processing-based method. Zhao et al. [30] proposed a method to estimate creep forces and friction coefficients using the unscented KF. Hussain et al. proposed a multiple model estimation approach for the identification of the adhesion limit in wheel–rail contact using a bank of KFs [31]. Mosleh et al. introduced a wheel flat detection method based on the continuous wavelet transform and a features extraction technique [32].

4.4.3 Track Condition Monitoring

Track irregularities are closely related to vehicle ride quality and safety and are one of the most important inspection items. Track irregularities include gauge, longitudinal level, alignment, cross level and twist irregularities.

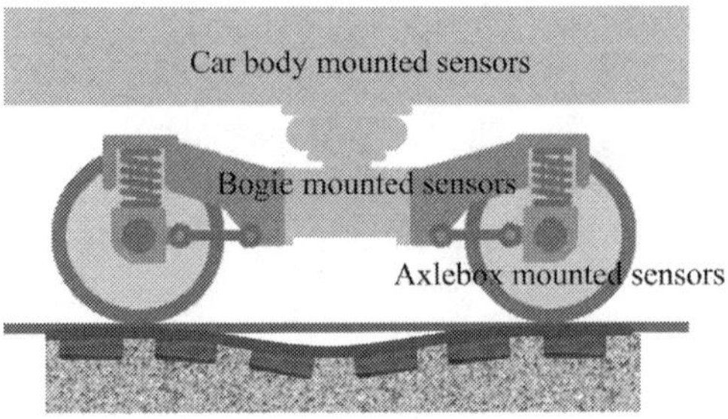

FIGURE 4.8 Sensor configuration for in-service vehicle.

There are three major types of sensor arrangements for track condition monitoring using in-service trains, as shown in Figure 4.8. The first is to mount the sensor on the axlebox.

Inertial measurement is a well-known method for measuring track displacement. This method has the advantage of direct measurements of the track condition. The inertial measurement method uses the physical law that displacement can be calculated by integrating acceleration twice and calculates track displacement from the output of an accelerometer attached to the axlebox. The relationship between axlebox accelerations and railway track defects or irregularities has been analysed and used to identify track faults [33–37].

The second method is to mount the sensor on the bogie. This method is easier to maintain than mounting on the axlebox but requires careful consideration of the position of the sensor on the bogie. A track irregularity monitoring method using bogie-mounted sensors was proposed in [38–40].

The method with the best maintainability is to install sensors inside the car body. In this method, vibrations generated by track geometry are transmitted to the car body via the primary and secondary suspensions. Therefore, it is necessary to examine the effect of track conditions on the car body vibration. Tsunashima et al. developed a system to identify track faults using accelerometers and a GNSS antenna placed on the car bodies of in-service vehicles [41, 42]. Bai et al. used low-cost accelerometers placed on or attached to the floors of operating trains to analyse track conditions [43]. Track condition monitoring based on bogie and car body acceleration measurements was presented and verified [44]. Balouchi presented a cab-based track-monitoring system [45]. Chellaswamy et al. described a method for monitoring railway track irregularities by updating the status of tracks in the cloud [46].

Recently, a method was proposed for diagnosing track conditions by substituting dedicated sensing devices with general-purpose smartphones. Rodríguez et al. proposed the use of mobile applications to assess the quality and comfort of railway tracks [47]. Cong et al. described the use of a smartphone as a sensing platform to obtain real-time data on vehicle acceleration, velocity and location [48]. Paixão et al. used smartphones to perform constant acceleration measurements inside in-service trains, which can complement the assessment of the structural performance and geometrical degradation of the tracks [49]. A smartphone-based track condition monitoring system was developed for regional railways in Japan [50].

Model-based estimation techniques have been proposed to solve the inverse problem of estimating track geometry from vehicle vibrations. KF-based methods have been developed to estimate the track geometry from car body motions [51, 52]. A nonlinear model-based estimation procedure, based on a central difference KF, was used to estimate the lateral wheel-rail contact forces and moments, including the identification of lateral track irregularities [53].

However, these methods present significant challenges for railway operators when they are used daily. Paglia et al. described a method to predict track longitudinal level irregularity using the bogie vertical acceleration from in-service vehicles [54]. A linear regression model was used to estimate track irregularity. Tsunashima proposed a method for estimating track irregularity using Gaussian Process Regression (GPR) based on data obtained from a vehicle travelling simulation [55, 56].

Time-frequency analysis can be applied for identifying track faults [57, 58]. Tsunashima et al. described a method for automatically classifying the type and degradation level of track using a convolutional neural network by imaging car body acceleration on a time-frequency plane by continuous wavelet transform [59]. A Wasserstein generative adversarial network-based framework was developed for real-time track irregularity estimation [60]. A deep learning technique was applied to car body acceleration for a high-speed railway for estimating track geometry [61]. A lightweight multi-layer-perceptron deep learning architecture was used to estimate track vertical irregularities of railway bridges from vehicle responses [62].

4.4.4 System Development for Conventional Lines

Condition monitoring is a technology that has been put to practical use in recent years due to the rapid development of sensor technology, and is based on the concept of maintaining a certain level of safety and other functions by constantly monitoring the condition of various parts of the railway system while it is in use. If the condition of a running vehicle can be measured from the ground, it is possible to monitor the condition of all vehicles passing a certain point, and conversely, if the condition of the track on the ground can be monitored from the vehicle side, it is possible to determine the condition of the track on which the vehicle is running.

4.4.4.1 PQ Monitoring Bogies

In underground railways, where there are many sharp curves, it is extremely important that the vehicles pass through the curves safely and smoothly. To achieve this, it is important to reduce the derailment coefficient by reducing the lateral force when passing through curves, to reduce noise and vibration associated with the passage, and to reduce unusual wear such as lateral wear of flanges and rails, as well as rail corrugation.

The derailment coefficient, which is used as an indicator of the running safety of railway vehicles, is the ratio P/Q of the longitudinal (wheel load) force P to the lateral force Q applied to the wheels, and the derailment coefficient that occurs while the train is running is lower than the critical derailment coefficient (an indication of safety) calculated from the contact geometry and friction coefficient between the wheels and rails. The safe running of a train is ensured if the derailment coefficient is lower than the derailment limit coefficient (the reference value for safety) calculated from the contact geometry and friction coefficient between wheels and rails.

Traditionally, derailment coefficients have only been measured on limited occasions, mainly when new lines are opened or new rolling stock is introduced, and there has been no observation of how they change over time. However, recent research has shown that the derailment coefficient changes even when the same vehicle passes the same point on the same route. This has led to a need for the development of a constant monitoring system on in-service trains to capture changes in derailment coefficients under different conditions.

Conventional derailment coefficient measurements require the use of special wheelsets called PQ wheelsets. These measure wheel load and lateral force from wheel strain and cannot be installed on operational trains due to the fact that they have many strain gauges attached and holes drilled to allow signal lines to pass through, and the brakes cannot be applied to avoid thermal effects, thus making it possible only on limited occasions to measure the derailment coefficient with these wheelsets on special trains. Measurements could only be made with a limited number of running tests. In addition, a slip ring or telemeter is required to retrieve signals from the rotating PQ wheelset, which is not suitable for constant monitoring. Therefore, a new method to measure wheel load and lateral force without using the PQ wheelset was developed, using sensors mounted on the bogie rather than on the wheels (Figure 4.9).

The wheel load was measured by capturing the deformation of the bogie instead of the wheel strain, and the vertical displacement of the springs in the axle was measured by a magneto-strictive sensor. For the measurement of lateral force, the deformation of the wheel was directly measured by a non-contact displacement transducer installed near the axlebox of the bogie. The wheel is not only deformed by lateral force while running, but also by lateral displacement due to rattling of the bearings and displacement due to the inclination of the wheel itself. To filter out these effects and to capture only the deformation caused by lateral force, a supplementary sensor was installed. The use of conical roller bearings and the

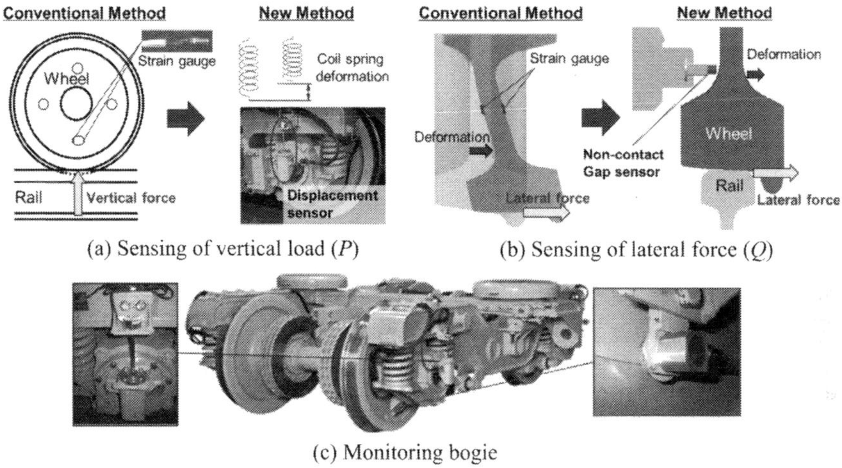

FIGURE 4.9 PQ monitoring bogie.

fixing of the bearings minimised lateral displacement as much as possible and made it possible to accurately detect small deformations caused by lateral forces.

Comparisons with measurements using conventional PQ wheelsets show that the measured values are in good agreement and that there are no practical problems. In addition, for constant monitoring on commercial trains, the brakes are not applied to the wheels but to brake discs to avoid thermal deformation of the wheels due to braking.

As a result of these various measures, bogies that can constantly measure the derailment coefficient on in-service trains have been produced and measurement data are obtained on a daily basis using in-service trains. These bogies are used on the Marunouchi, Tozai and Chiyoda lines of the Tokyo Metro system as well as on the Shinjuku Line of the Tokyo Metropolitan Transportation Bureau in Japan.

4.4.4.2 Track Monitoring Equipment

The track monitoring system, developed by JR East, Japan Railway Track Consultants and others, monitors track irregularities and the condition of sleepers, rail fasteners and other parts of the track from in-service trains, and is mounted under the floor of the train to enable high-frequency data acquisition (Figure 4.10).

Track monitoring equipment consists of two devices: 'Track displacement monitoring equipment' and 'Track material monitoring equipment'. Track displacement monitoring equipment acquires track displacement data using laser displacement sensors, gyroscopes and accelerometers, while track material monitoring equipment captures images of the state of components such as sleepers and rail fasteners, and uses the image data to automatically or visually identify faults in the track.

FIGURE 4.10 Track monitoring system developed by JR East.

4.4.4.3 Wheel Flat Detector

Flattening, in which part of the wheel is shaved off and the wheel's roundness is impaired due to sliding during braking, is a lingering problem that causes significant noise and vibration, so it must be detected and counteracted as soon as possible. For this reason, various detection systems have been developed and are used by railway operators. The system measures and analyses the vibration and noise of a passing train from the track-side. Generally, the system consists of an accelerometer to detect the impact acceleration caused by flats and a wheel detector to detect the passing of the wheels.

4.4.4.4 Wheel Load Measuring Device

In the Hibiya Line Nakameguro accident in 2000[1], one of the reasons behind the accident was the imbalance between the left and right wheel loads, which resulted in the Japan MLIT issuing a notification to control the left-right wheel load imbalance to 10% or less. For this reason, railway operators have considered ways to accurately measure the load of each wheel in a bogie, and equipment has been developed and installed to measure the wheel load of wheelsets in service. Such systems use load cells or strain gauges affixed to the rails to measure shear or bending strain.

Figure 4.11 shows the principal of measuring lateral and vertical forces with strain gauges attached to the rail, as introduced in Tokyo Metro. In this figure, the wheel between sleepers provides a vertical force (V) on to the rail and the forces W_1 and W_2 react from sleepers to the rail. These three forces acting on the rail produce shearing strains, which are drawn with arrows on the surface of the rail web. These shearing strains are proportional to the vertical force. Figure 4.11(b) shows the

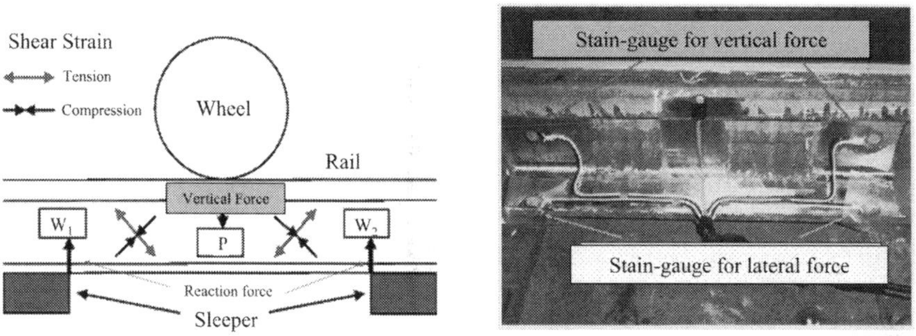

(a) Principle of measuring vertical force (b) Strain-gauges attached to the rail

FIGURE 4.11 Wheel load and lateral force measurement device.

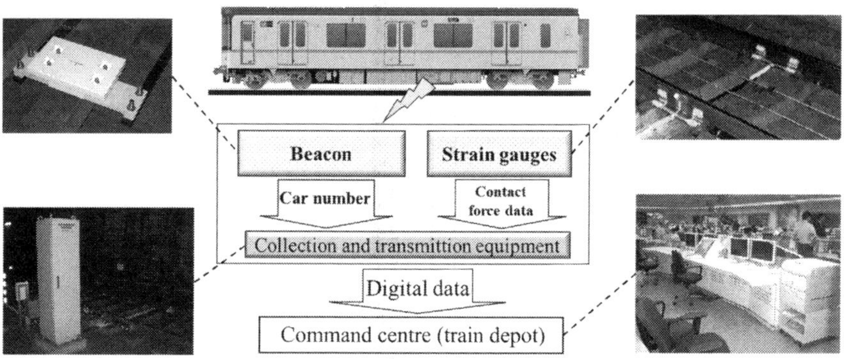

FIGURE 4.12 Realtime PQ Monitoring system from the track-side.

strain gauges attached to the rail. The principal of measuring lateral force is the same as that of vertical force.

Figure 4.12 shows the overall monitoring system. The sensors, which measure lateral and vertical forces using strain gauges, are attached to the rail. The real-time monitoring equipment consists mainly of data acquisition equipment located at specific curves on a commercial line and monitoring equipment located at train depot offices and elsewhere. Data collection equipment begins to measure the lateral force and the vertical force and automatically detects the passing of vehicles at the measurement point. Specifically, it gathers several types of data while a train passes through that point. Train information obtained via a beacon (such as the number of operations, number of trains and operating company, taking into account interconnections between different railway companies) is attached to the measured data. The data collection equipment automatically transmits these data to monitoring equipment at the depot, office and command centre via a fibre optic network. The monitoring equipment receives data each time the wheels pass the measurement point and analyses the wave form data for specific values that indicate the performance of the vehicle dynamics.

A series of freight train derailments on the JR Hokkaido Esashi Line between 2012 and 2014 was investigated by the Japan Transport Safety Board, and a recommendation was issued to the MLIT regarding improvements to the imbalance in wheel load of freight cars due to uneven container loading. The Japan Freight Railway Company has introduced a device to check the wheel load imbalance of wagons in service. This method was adopted because it is extremely difficult to check the internal condition of containers during transport.

4.4.4.5 Derailment Detector

There have been accidents in which a car at the rear of the train derailed, but the driver was unaware of it and continued to drive the train, causing the accident to escalate in severity. If the driver is informed of a derailment as it unfolds, the driver can take actions such as emergency braking and the accident can be prevented from growing more severe. In Japan, Tokyu Car Corporation (now Japan Transport Engineering Company) and JR East jointly developed a type of acceleration sensor mounted to car bodies at both ends of the train. After adding and improving algorithms to detect overturning and collisions as well as derailments, the system has been installed on all of JR East's E331 series trains since 2006. Subsequently, this system was improved and developed, and a type with sensors in all cars, including the intermediate cars, began to be installed on the Tokyo Metro's Hibiya Line 13000 Series from December 2018.

JR West has been equipped with an unusual train behaviour detection system for Series 521 trains (3rd generation) since they were introduced in FY2013. The system is designed to perform emergency braking when the value reported by the acceleration sensor installed on the car body frame near the bogie of each car exceeds a threshold that takes the possibility of derailment, overturn or collision into account.

4.4.5 System Development in Shinkansens

Shinkansens carry out daily transport using a large number of vehicles and other assets and systems. Monitoring these assets and systems is an important aspect of operation. Using dedicated vehicles to collect data on assets along the line (for example, track, overhead contact line, signalling and radio) has been applied since the beginning of the Shinkansen service. Data collection methods for track-side assets using in-service trains are also being developed by JR companies, including JR Kyushu.

4.4.5.1 Monitoring of Track-Side Assets Using Electric and Track Inspection Cars

(1) Tokaido and Sanyo Shinkansens: An electric and track inspection car (three-bogie system), capable of travelling at 210 km/h, was completed in 1974.

Along with various channels of data collection, it also monitored the power collection condition of the pantograph systems from the observation dome. In 2001, a car based on the Series 700 Shinkansen train (two-car system) was introduced. On the Tokaido and Sanyo Shinkansen lines, these electric and track inspection cars (known as "Doctor Yellow") are used to carry out measurements on the entire line at approximately 10-day intervals.

(2) Tohoku, Joetsu and Hokuriku Shinkansens: In 1982, a modified Type 925 Shinkansen test train was used as an electric and track inspection car. Later, in 2001, track electric and track inspection cars Type E926 (East Eye), based on Series E3 Shinkansen trains (six-car train), were used. These vehicles enable on-board data analysis and easy communication with the command centre, allowing rapid action to be taken depending on the data provided. The acquired data is analysed in detail, including time series, by each maintenance department and used to plan maintenance work.

4.4.5.2 Monitoring of Track-Side Assets Using In-Service Trains

Electric and track inspection cars are not used on Kyushu Shinkansen due to the line conditions and the efficiency of maintenance work. Since the start of operations, measurement equipment has been installed on the roof of the 800 Series Shinkansen trains to collect data during in-service runs. For the track, data were obtained by running test trains at 120 km/h during nighttime hours; in 2010, dedicated measuring equipment was developed for in-service trains. This was mounted on the lower part of the 800 Series bogie and measurements were commenced. Further developments have also made it possible to measure the wear of contact wires in a catenary system at high speeds using an in-service train.

On the Tokaido Shinkansen, equipment for automatically measuring vertical rail changes was installed on Series 300 Shinkansen trains from 2002, and data collected on in-service trains were recorded on IC cards and analysed at the depot. Today, in-service trains can be used to measure track irregularities, which are mounted on N700S trains and the measurement data is transmitted to the depot using leaky coaxial cable radio transmission.

4.4.5.3 Track-Side-Based Vehicle Condition Monitoring

As a method for monitoring passing vehicles from the track-side, axlebox heat detection systems have been used on the Tokaido Shinkansen since its inauguration (Figure 4.13). Temperature sensors are installed alongside the tracks to monitor the heat generated in the axleboxes of passing trains, and an alarm is used to alert the relevant authorities in the event of a malfunction. Improvements have been made to this method, such as increasing the accuracy of measurement and reading the train number. A method of measuring by infrared sensors has been in use since 2007. A method for detecting temperature rises in the axlebox and related components around the wheelset from the underside and sides of the track began operating on the Tokaido Shinkansen in 2015, with the number of measuring points increased to

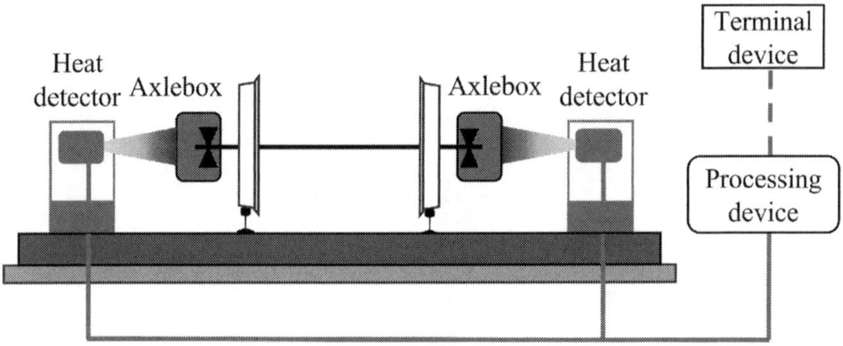

FIGURE 4.13 Axlebox heat detector developed by JR Central.

five in 2020. This has made it possible to monitor temperature changes in the same part of the train during running.

4.5 CONDITION-BASED MAINTENANCE (CBM)

Railways consist of many subsystems, including structures (track), mechanical systems (wheels, bogies, and car bodies), vehicle traction and braking systems (motors and their control units, brakes and their control units), power supply systems (substations and power cables), signalling and security systems (on-ground and onboard), communication systems and other systems (for example, stations and level crossings). The basic principle has been to inspect and repair each subsystem to maintain a certain level of performance, and regulations are also based on this concept.

Time-based maintenance is relatively easy to manage and provides a high degree of reliability, but it also carries with it the potential for unnecessary maintenance and the inability to respond to rapid changes. In contrast, CBM enables flexible maintenance in accordance with the state of degradation, and maintains a certain level of safety and other performance metrics at a lower cost. These methods have become possible largely due to improvements in sensor technology that have made it relatively easy to monitor the condition of vehicles and tracks.

If the condition of a travelling vehicle can be measured from the track, it is possible to monitor the condition of the wheels and other components of all vehicles passing a certain point, and if the condition of the track can be monitored from the vehicles, it is possible to monitor the condition of the line the vehicle is travelling on. Such monitoring is in high demand in Europe, where freight wagons and other vehicles of many nations operate across national borders, and has been developed and put into practical use in Japan. CBM is not only effective in reducing costs and manpower, but is also capable of maintaining effective and safe conditions, and therefore inspection and maintenance is expected to move towards this approach in the future.

4.5.1 Achieving Good Curve Passage with CBM

CBM, in which track and vehicle maintenance is performed while continuously monitoring wheel/rail conditions, has proved very effective in solving safety problems such as reducing the increase in derailment coefficient and maintenance problems such as reducing rail corrugation and rail/wheel side wear. These wear conditions are detrimental and especially occur on curved sections. CBM is very effective in solving maintenance problems such as rail corrugation wear and reduction of rail-wheel side wear.

Figure 4.14 illustrates conceptually how CBM can improve curve passing performance and at the same time prevent undesirable wheel–rail phenomena. First, it is important that the measures to achieve perfect rolling on sharp curves are reflected as basic specifications in the design, manufacturing and installation stages for bogies, other rolling stock, rails and other track. In other words, these must first be considered in the basic concept. Next, in the actual operation phase, the condition of the wheel-rail contact system is continuously monitored, and based on the results, areas that cannot be covered by the basic design specifications or that cannot be achieved due to deterioration of the basic specifications are supplemented or corrected by oil lubrication, friction adjustment or wheel and/or rail shaving, to maintain the wheel–rail system in good condition. In other words, CBM is important for maintaining the wheel and rail system in good condition.

The condition of the wheel–rail contact system as it passes through curves changes from time to time, and on-board monitoring using PQ monitoring bogies and other commercial trains, as well as ground monitoring at points where necessary, can help control friction, including the amount and timing of lubricant and friction modifiers, and maintain wheel tread and rail head profiles. It is also

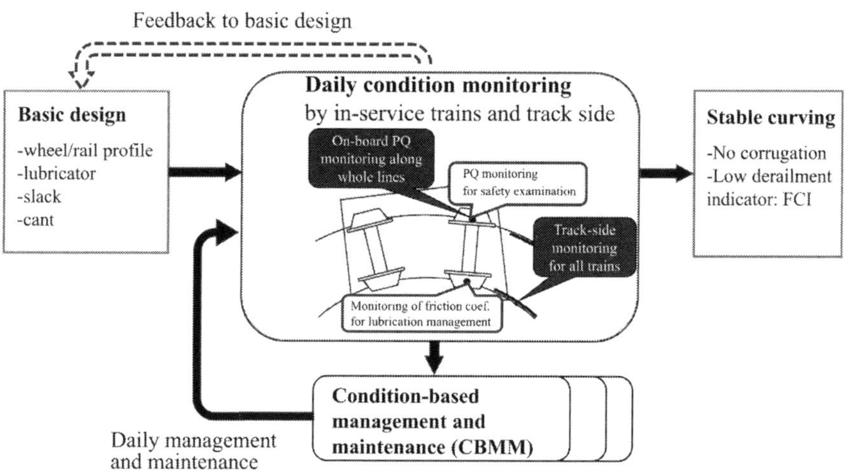

FIGURE 4.14 Curving performance improvement by CBM.

important to maintain the wheel tread profile and rail top profile by grinding and other measures to ensure proper vehicle maintenance and track repair.

These measures help to eliminate harmful phenomena on curves, such as rail corrugation, while maintaining safe train operation with a low derailment coefficient, thereby achieving safe curve passing conditions. In addition, constant monitoring can contribute to safer curve passing by analysing the big data obtained, which can be fed back to improve the basic specifications.

NOTE

1 The train disaster occurred in Japan on 8 March 2000 in Japan. A commuter train derailed on the transient curve section of 160.1m radius curve, and crashed against the opposite-direction train, and 5 passengers were killed. It was the typical flange-climb derailment at low speed caused by multiple factors.

REFERENCES

1. R.J. Patton, P.M. Frank R.N. Clark, *Issues of Fault Diagnosis for Dynamic System*, Springer, 2000.
2. S. Bruni, R. Goodall, T.X. Mei, H. Tsunashima, Control and monitoring for railway vehicle dynamics, *Vehicle System Dynamics*, 45, 743–799, 2007.
3. R.M. Goodall, C. Roberts, Concept and techniques for railway condition monitoring, *Proceedings of 2006 IET International Conference On Railway Condition Monitoring*, 90–95, 2006.
4. S. Bruni, R. Goodall, T.X. Mei, H. Tsunashima, Control and monitoring for railway vehicle dynamics, *Vehicle System Dynamics*, 45, 743–799, 2007.
5. C. Li, S. Luo, C. Cole, M. Spiryagin, An overview: Modern techniques for railway vehicle on-board health monitoring systems, *Vehicle System Dynamics*, 55(7), 1045–1070, 2017.
6. G. Charles, R. Goodall, R. Dixon, Model-based condition monitoring at the wheel–rail interface, *Vehicle System Dynamics*, 46(sup1), 415–430, 2008.
7. P. Li, R. Goodall, V. Kadirkamanathan, Estimation of parameter in a linear state space model using a Rao-Blackwellised particle filter, *IEE Proceedings-Control Theory Applications*, 151(6), 727–738, 2004.
8. S. Alfi, S. Bruni, L. Mazzola, S. Bionda, Model-based fault detection in railway bogies, In: Iwnicki, S., Goodall, R., Mei, T.X. (eds) *Proceedings of the 22nd IAVSD Symposium* , Manchester Metropolitan University, 2011.
9. X.Y. Liu, S. Alfi, S. Bruni, An efficient recursive least square-based condition monitoring approach for a rail vehicle suspension system, *Vehicle System Dynamics*, 54(6), 814–830, 2016.
10. M. Jesussek, K. Ellermann, Fault detection and isolation for a nonlinear railway vehicle suspension with a Hybrid Extended Kalman filter, *Vehicle System Dynamics*, 51(10), 1489–1501, 2013.
11. Hayashi, H. Tsunashima, Y. Marumo, Fault detection of railway vehicle suspension systems using multiple-model approach, *Journal of Mechanical Systems for Transportation and Logistics*, 1(1), 88–99, 2008.

12. H . Mori, H. Tsunashima, Condition monitoring of rail- way vehicle suspension using multiple model approach, *Journal Mechanical Systems Transport Logistics*, 2010(3), 154–165, 2010.
13. M. Jesussek, K. Ellermann, Fault detection and isolation for a full-scale railway vehicle suspension with multiple Kalman filters, *Vehicle System Dynamics*, 52(12), 1695–1715, 2014.
14. S. Zoljic-Beglerovic, B. Luber, G. Stettinger, G. Muller, M. Horn, Parameter identification for railway suspension systems using Cubature Kalman Filter, In: Klomp, M., Bruzelius, F., Nielsen, J., Hillemyr, A. (eds) *Advances in Dynamics of Vehicles on Roads and Tracks. IAVSD 2019. Lecture Notes in Mechanical Engineering*. Springer, 2020.
15. F. Liu, H. Zhang, X. He, Y. Zhao, F. Gu, A.D. Ball, Correlation signal subset-based stochastic subspace identification for an online identification of railway vehicle suspension systems, *Vehicle System Dynamics*, 58(4), 569–589, 2019.
16. D. Lebel, C. Soize, C. Funfschilling, G. Perrin, High-speed train suspension health monitoring using computational dynamics and acceleration measurements, *Vehicle System Dynamics*, 58(6), 911–932, 2019.
17. G. Vlachospyros, D. Fassois, J.S. Sakellariou, On-board vibration-based robust and unsupervised degradation detection in railway suspensions under various travelling speeds via a Multiple Model framework, *Vehicle System Dynamics*, 62(6), 1446–1470, 2023.
18. R. Ghiasi, M. Gordan, A. Mosleh, D. Ribeiro, A. Malekjafarian, M-CLUSTER: Multistage clustering for unsupervised train wheel condition monitoring, *Vehicle System Dynamics*, Published online: 12 Nov 2024, 1–26, 2024.
19. J. Sun, E. Meli, X. Song, M. Chi, W. Jiao, Y. Jiang, A novel measuring system for high-speed railway vehicles hunting monitoring able to predict wheelset motion and wheel/rail contact characteristics, *Vehicle System Dynamics*, 61(6), 1621–1643, 2022.
20. A. Matsumoto, Y. Michitsuji, Y. Ichiyanagi, Y. Sato, H. Ohno M. Tanimoto, A. Iwamoto, T. Nakai, Safety measures against flange-climb derailment in sharp curve -considering friction coefficient between wheel and rail-, *Wear*, 432–433, 202931, 2019.
21. F. Xia, C. Cole, P. Wolfs, A method for setting wagon speed restrictions based on wagon responses, *Vehicle System Dynamics Supplement*, 44, 424–432, 2006.
22. A. Matsumoto, Y. Sato, H. Ohno, M. Tomeoka, K. Matumoto, J. Kurihara, T. Ogino, M. Tanimoto, Y. Kishimoto, Y. Sato, T. Nakai, A new monitoring method of train derailment coefficient, *International Conference on Railway Condition Monitoring*, 2006, 136–140, 2006.
23. A. Iwamoto, K. Matsumoto, K. Yano, T. Fukushima, A. Mat- sumoto, Y. Sato, H. Ohno, Y. Michitsuji, M. Tanimoto, D. Shinagawa, Monitoring of Derailment coefficients Y/Q for in-service trains and the analysis for the increase of curve safety, *Proceedings of the Third International Conference on Railway Technology: Research, Development and Maintenance, Paper No. 64*, 2006.
24. T. Fukushima, K. Yano, A. Iwamoto, K. Matsumoto, A. Matsumoto, M. Tanimoto, Y. Michitsuji, Y. Sato, M. Mizuno, D. Shinagawa: Derailment coefficient data for commercial lines measured by PQ monitoring bogies and methods of application, *Proceedings of International Symposium on Speed-up and Sustainable Technology*

for Railway and Maglev Systems 2015 (STECH2015), Japan Society of Mechanical Engineers, November 10-12, 2015.

25. Y. Boronenko, A. Orlova, A. Iofan, S. Galperin, Effects that appear during the derailment of one wheelset in the freight wagon – simulation and testing, *Vehicle System Dynamics Supplement*, 44, 663–668, 2006.
26. W. Hubacher, P. Scheiber, Derailment detector systems for CFF/SBB rolling stock, *Rail International*, 31, 31–37, 2000.
27. S. Koga, K. Miyakawa, M. Kageyama, N. Shimada, Development of a vehicle abnormal behavior detection system, *Proceedings of the Third International Conference on Railway Technology: Research, Development and Maintenance, J. Pombo (Editor), Civil-Comp Press, Paper* No. 78, 2016.
28. C.P. Ward, R.M. Goodall, R. Dixon, G.A. Charles, Adhesion estimation at the wheel–rail interface using advanced model-based filtering, *Vehicle System Dynamics*, 50(12), 1797–1816, 2012.
29. P. Hubbard, C. Ward, R. Dixon, R. Goodall, Real time detection of low adhesion in the wheel/rail contact, *Proceedings of IMechE Part F: Journal of Rail and Rapid Transit*, 227(6), 623–634, 2013.
30. Y. Zhao, B. Liang, S. Iwnicki, Friction coefficient estimation using an unscented Kalman filter, *Vehicle System Dynamics*, 52(supl), 220–234, 2014.
31. I. Hussain, T.X. Mei, R.T. Ritchings, Estimation of wheel–rail contact conditions and adhesion using the multiple model approach, *Vehicle System Dynamics*, 51(1), 32–53, 2012.
32. A. Mosleh, A. Meixedo, D. Ribeiro, P. Montenegro, R. Calçada, Early wheel flat detection: an automatic data-driven wavelet-based approach for railways, *Vehicle System Dynamics*, 61(6), 1644–1673, 2022.
33. X. Chen, X. Chai, X. Cao, The time-frequency analysis of the train axle box acceleration signals using empirical mode decomposition, *Computing Model New Technology*, 18, 356–360, 2014.
34. T. Karis, M. Berg, S. Stichel, Correlation of track irregularities and vehicle responses based on measured data, *Vehicle System Dynamics*, 56, 967–981, 2018.
35. H.C. Tsai, C.Y. Wang, N.E. Huang, Railway track inspection based on the vibration response to a scheduled train and the Hilbert–Huang transform, *Proceedings Institution Mechanical Engineering, Part F: Journal of Rail and Rapid Transit*, 229, 815–829, 2014.
36. X. Sun, Y. Fei, J. Shi, K. Zaitian, Z. Yunlai, On-board detection of longitudinal track irregularity, *IEEE Access*, 9, 14025–14037, 2021.
37. A. Chudzikiewicz, R. Bogacz, M. Kostrzewski, R. Konowrocki, Condition monitoring of railway track systems by using acceleration signals on wheelset axle-boxes, *Transport*, 33, 555–566, 2019.
38. P.F. Weston,. C.S. Ling, C. Roberts, C.J. Goodman, P. Li, R.M. Goodall, Monitoring vertical track irregularity from in-service railway vehicles, *Proceedings Institution Mechanical Engineering, Part F: Journal of Rail and Rapid Transit*, 221, 75–88, 2007.
39. P.F. Weston,. C.S. Ling, C. Roberts, C.J. Goodman, P. Li, R.M. Goodall, Monitoring lateral track irregularity from in-service railway vehicles, *Proceedings Institution Mechanical Engineering, Part F: Journal of Rail and Rapid Transit*, 221, 89–100, 2007.

40. A. Malekjafarian, E. Obrien, P. Quirke, C. Bowe, Railway track monitoring using train measurements: An experimental case study, *Applied Sciences*, 9, 4859, 2019.
41. H. Tsunashima, H. Ono, T. Takata, S. Ogata, Development and operation of track condition monitoring system using in-service train, *Applied Sciences*, 13, 3835, 2023.
42. H. Ono, H. Tsunashima, T. Takata, S. Ogata, Development and operation of a system for diagnosing the condition of regional railways tracks, *Mechanical Engineering Journal*, 10, 22–00239, 2023.
43. L. Bai, R. Liu, Q. Li, Data-driven bias correction and defect diagnosis model for in-service vehicle acceleration measurements, *Sensors*, 20, 872, 2020.
44. X. Wei, F. Liu, L. Jia, Urban rail track condition monitoring based on in-service vehicle acceleration measurements, *Measurement*, 80, 21–228, 2016.
45. F. Balouchi, A. Bevan, R. Formston, Development of railway track condition monitoring from multi-train in-service vehicles, *Vehicle System Dynamics*, 59, 1397–1417, 2021.
46. C. Chellaswamy, T.S. Geetha, A. Vanathi, K. Venkatachalam, An IoT based rail track condition monitoring and derailment prevention system, *International Journal of Technology*, 11, 81–107, 2020.
47. A. Rodríguez, S. Sanudo, M. Miranda, A. Gomez, J. Benavente, Smartphones and tablets applications in railways, ride comfort and track quality, *Transition Zones Analysis, Measurement*, 182, 109644, 2021.
48. J. Cong, M. Gao, M. Miranda, Y. Wang, R. Chen, P. Wang, Subway rail transit monitoring by built-in sensor platform of smartphone, *Frontiers of Information Technology & Electronics Engineering*, 21, 1226–1238, 2020.
49. A. Paixão, E. Fortunato, R. Calçada, Smartphone's sensing capabilities for on-board railway track monitoring, structural performance and geometrical degradation assessment, *Advances Civil Engineering,* 2019, 1729153, 2019.
50. H. Tsunashima, R. Honda, A. Matsumoto, Track condition monitoring based on in-service train vibration data using smartphones, In: Hessami, G., Muttram, R., (eds) *Civil Engineering*, InTech, 1–18, 2024.
51. M. Odashima, S.Azami, Y. Naganuma, H. Mori, H. Tsunahsima, Track geometry estimation of a conventional railway from car-body acceleration measurement, *Mechanical Engineering Journal*, 4, 16–00498, 2017.
52. X. Xiao, Z. Sun, W. Shen, A Kalman filter algorithm for identifying track irregularities of railway bridges using vehicle dynamic responses, *Mechanical Systems Signal Processing*, 138, 106582, 2020.
53. S. Strano, M. Terzo, C. Tordela, Output-only estimation of lateral wheel-rail contact forces and track irregularities, *Vehicle System Dynamics*, 62(10), 2481–2509, 2023.
54. L. Paglia, A. Reyes, D. Gialleonar, A. Facchinetti, M. Carnevale, A speed-dependent condition monitoring system for track geometry estimation using inertial measurements, In: Huang, W., Ahmadian, M. (eds) *Advances in Dynamics of Vehicles on Roads and Tracks III. IAVSD 2023. Lecture Notes in Mechanical Engineering*, Springer, 2025.
55. H. Tsunashima, Data-driven track irregularity estimation technique using car-body vibration, In: Huang, W., Ahmadian, M. (eds) *Advances in Dynamics of Vehicles on Roads and Tracks III. IAVSD 2023. Lecture Notes in Mechanical Engineering*, Springer, 2025.

56. H. Tsunashima, N. Yagura, Railway track irregularity estimation using car body vibration: A data-driven approach for regional railway, *Vibration*, 7, 928–948, 2024.
57. H. Tsunashima, R. Hirose, Condition monitoring of railway track from car-body vibration using time–frequency analysis, *Vehicle System Dynamics*, 60(4), 1170–1187, 2020.
58. X. Wei, X. Yin, Y. Hu, Y. He, L. Jia, Squats and corrugation detection of railway track based on time-frequency analysis by using bogie acceleration measurements, *Vehicle System Dynamics*, 58(8), 1167–1188, 2019.
59. H. Tsunashima, M. Takikawa, Monitoring the condition of railway tracks using a convolutional neural network, In: Bulnes, F. (ed) *Recent Advances in Wavelet Transforms and Their Applications*, IntechOpen, 2022.
60. Z. Yuan, J. Luo, S. Zhu, W. Zhai, A Wasserstein generative adversarial network-based approach for real-time track irregularity estimation using vehicle dynamic responses, *Vehicle System Dynamics*, 60(12), 4186–4205, 2021.
61. X. Hao, J. Yang, F. Yang, X. Sun, Y. Hou, J. Wang, Track geometry estimation from vehicle–body acceleration for high-speed railway using deep learning technique, *Vehicle System Dynamics*, 61(1), 239–259, 2022.
62. B. Wang, Y. Zeng, D. Feng, J. Li, Track vertical irregularity estimation for railway bridges using a novel and lightweight deep-learning architecture, *Vehicle System Dynamics*, Published online: 09 Jul 2024, 1–28, 2024.

5 Application Examples

5.1 *Q/P* MONITORING IN URBAN RAILWAYS

5.1.1 Development of Continuous *Q/P* Monitoring System for In-Service Trains

The derailment coefficient (see Chapter 3) changes daily due to changes in the coefficient of friction between the wheels and the rails. To capture these changes, it is necessary to measure the PQ of the rolling stock on the operational line. Previous methods using 'instrumented wheels' with strain gauges and slip-rings or telemeters were difficult to handle and had poor durability. Research and development therefore began on a new measurement system without strain gauges and data transmission devices, using gap sensors on a bogie frame, which could be implemented on a commercial service train. The new system, called a 'PQ monitoring bogie', was successfully developed as shown in Figure 5.1 [1]. Trains with PQ monitoring bogies have been running daily on three Tokyo Metro lines as commercial trains, collecting large amounts of actual rail–wheel contact data.

The actual condition data collected by the PQ monitoring bogies has been very useful and important in influencing the implementation of condition-based maintenance of track and vehicles in the Tokyo Metro system. Figure 5.2(a) shows the outside rail Q/P values for each pass of a train on a single curve with a radius of 160 m. As shown in the figure, the value of Q/P changes each time a train passes, but the waveform is similar. Figure 5.2(b) shows the relationship between the value of Q/P and the value of inside Q/P, denoted by κ. Each data point is defined at the maximum value of Q/P in a curve. This figure shows that the value of Q/P is almost proportional to the value of κ and proves that both Q/P and κ vary widely during commercial train operation.

As shown in Figure 5.3, rail lubricators are attached at the start of sharp curves. Typically, the outer rail is lubricated from the Gauge Corner (GC) to prevent flange wear and the inner rail is lubricated from the Field Corner (FC) to reduce the lateral turning force. Because the direction of lubrication on the rails and the contact point between the wheel and the rail head are different, the friction conditions of the four wheels are also different. As a result, the difference in friction coefficients between

DOI: 10.1201/9781003362135-5

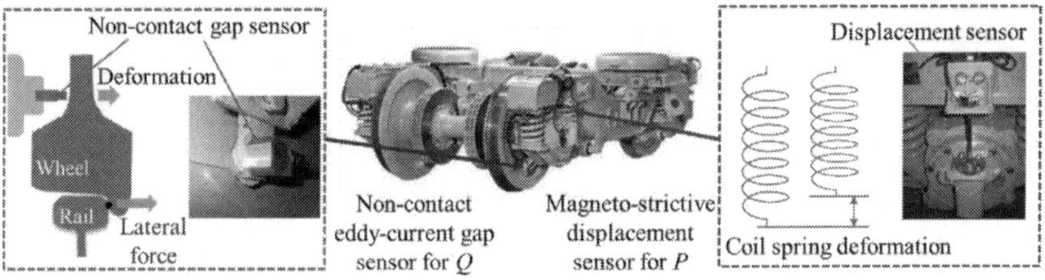

FIGURE 5.1 Measurement mechanism for PQ monitoring bogie.

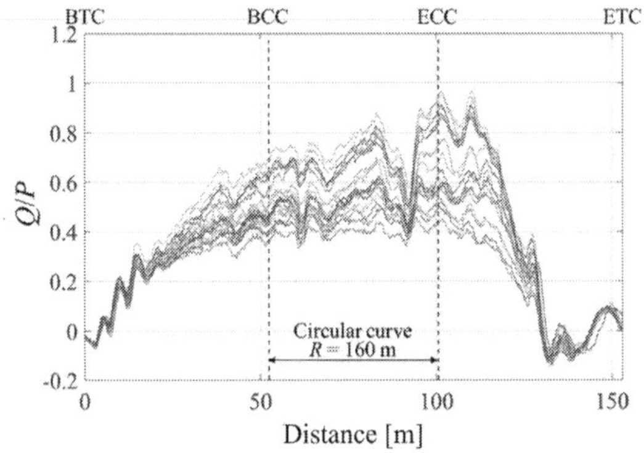

(a) Outside Q/P measured for each time pass of a train passes ($N = 16$).

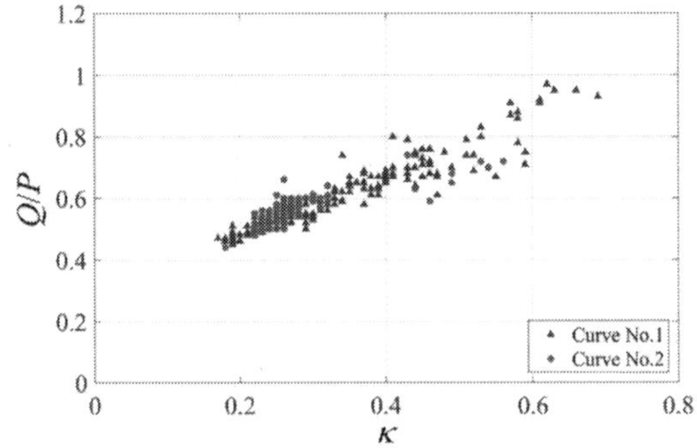

(b) Relationship between maximum value of Q/P and corresponding κ.

FIGURE 5.2 Actual states of Q/P in on a sharp curve measured by a PQ monitoring bogie.

the four wheels has a great influence on cornering dynamics and flange climbing avoidance.

Figure 5.4(a) shows the relationship between the derailment coefficient Q/P and the radius of curvature along the whole line. The value of Q/P increases with

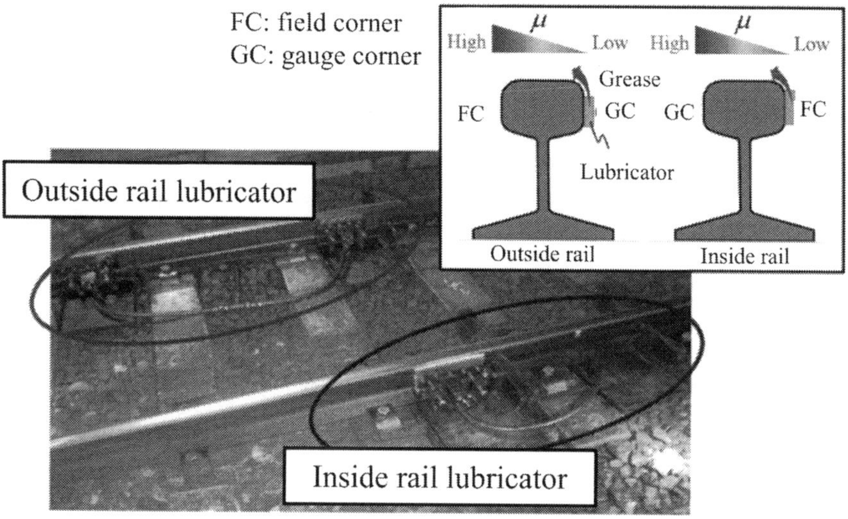

FIGURE 5.3 Rail lubrication devices attached to the beginning of a curve.

increasing curvature, as does the friction of the inner wheel–rail contact surface, which is influenced by the application of wayside lubrication. Figure 5.4(b) shows the relationship only on well-maintained curves. The Q/P values are lower than those for whole curves, and for curves with a radius greater than 200 m, the Q/P remains below the safety limit calculated by Nadal's formula. Tracks on curves should be maintained to keep Q/P below the safety limit. On curves with a radius of 200 m or less, anti-derailment guards should be used for safety. In this way, track maintenance based on continuous monitoring by PQ monitoring bogies can keep tracks in a safer condition.

5.1.2 Method for Measuring Twist Irregularity with a PQ Monitoring Bogie

The variation in the derailment coefficient Q/P is closely related to the variation in the wheel load of the bogie, which can be quantified by the twist irregularity. In this section, a method for estimating twist irregularities using a PQ monitoring bogie [2] is explained.

Figure 5.5 shows the wheel load P on a single curve. As shown in the figure, the twist irregularity of a track greatly influences the fluctuation of wheel load, especially in transition curves. Vertical displacements collected by the PQ monitoring bogie between the four axleboxes and the bogie frame are used to estimate the twist irregularity. Each measured displacement is denoted in Figure 5.6. The value of bogie twist T, which is closely related to the twist irregularity of the track, is described using the following equation:

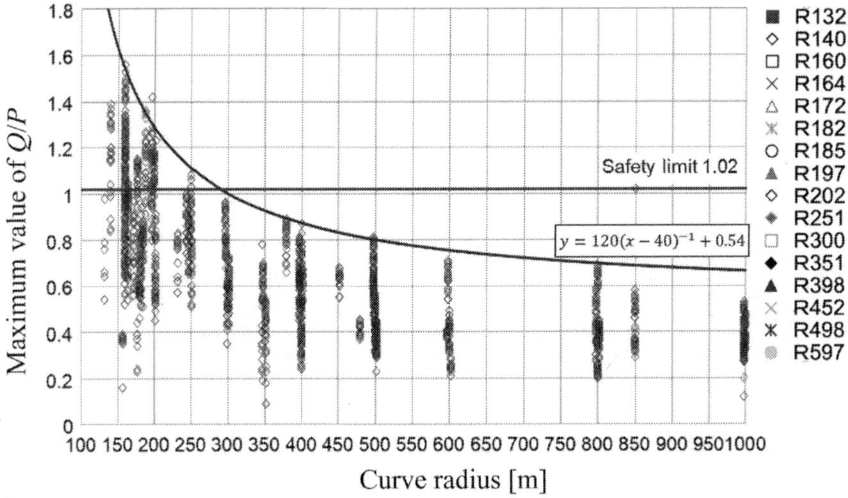

(a) On all curves

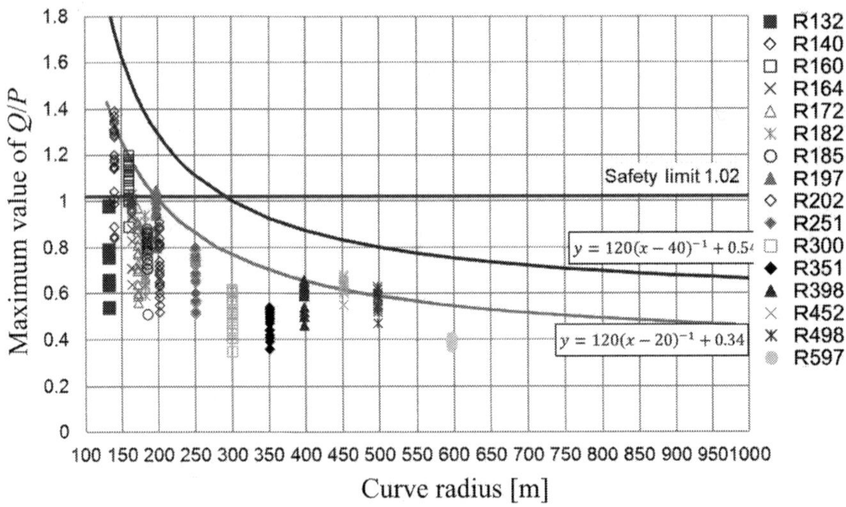

(b) On well-maintained curves

FIGURE 5.4 Relationship between derailment coefficients and curve radius [15].

$$T = \frac{b}{b_1}\{\left(z_{s1out} + z_{s2in}\right) - \left(z_{s1in} + z_{s2out}\right). \tag{5.1}$$

The calculated results using this equation are shown in Figure 5.7. Even though the value of wheel load widely changes as shown in Figure 5.5, the values of T do not change so much. This proves that the estimated value of T is almost invariant

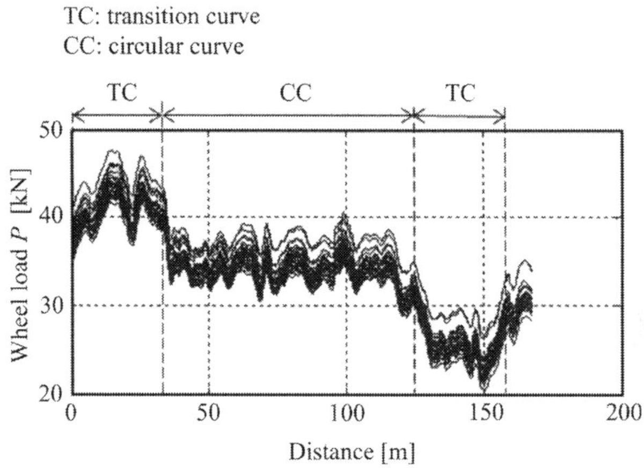

FIGURE 5.5 Wheel load of the leading outside wheel of the bogie (N = 35).

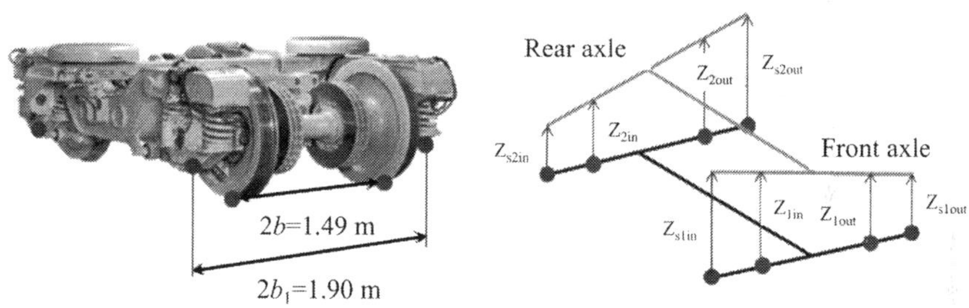

FIGURE 5.6 Image of the bogie twist in the monitoring bogie.

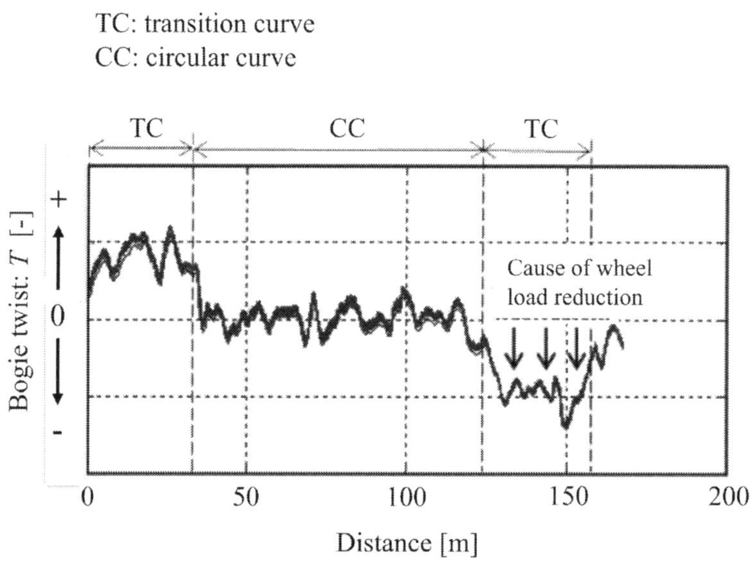

FIGURE 5.7 Calculated bogie twist values (N = 35).

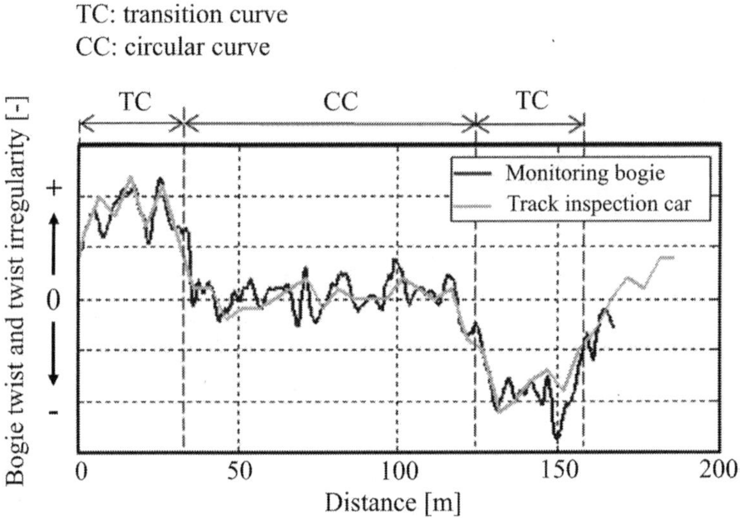

FIGURE 5.8 Comparison between the twist irregularity with a track inspection car and the bogie twist of the monitoring bogie.

for each train pass and shows high duplicability. Due to the cancelation effect of each displacement, the value is not affected by the passenger load factor change.

To evaluate the obtained bogie twist with the twist irregularity of the track, the value measured by a track inspection car is compared with the value measured by PQ monitoring bogie. Figure 5.8 shows the results of this comparison. The amplitude of the measured irregularity with a track inspection car is multiplied by 1.9/2.5 so that the difference in wheel bases, of which the monitoring bogie is 1.9 m and the track inspection car is 2.5 m, can be fairly compared. As shown in Figure 5.8, both waveforms correspond well to each other throughout the curve. Because the monitoring bogie produces a measurement every 0.5 ms, the resolution along the track is 1 cm when the vehicle is running at 20 m/s.

5.1.3 Temporal Subtraction Analysis Using a PQ Monitoring Bogie

The temporal subtraction processing of derailment coefficients is conducted using the previously described twist irregularity estimates. Figure 5.9 shows the waveforms after correction. As shown in the figure, both periods of the datasets seem to be corrected along the running distance whereas the amplitude shows fluctuations.

Figure 5.10 shows the average values in 2011 and 2012. As can be seen in the figures, the average tendency shows a difference. To clarify the tendency of running safety, Figure 5.11 shows the temporal subtraction of derailment coefficients. As shown in the figure, the value of derailment coefficient decreases to a maximum of 0.4 due to the application of lubricant from ground equipment. In this way,

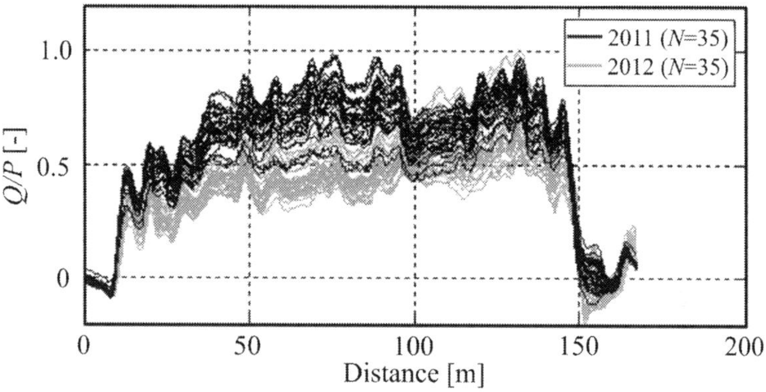

FIGURE 5.9 Comparison of derailment coefficients after correction.

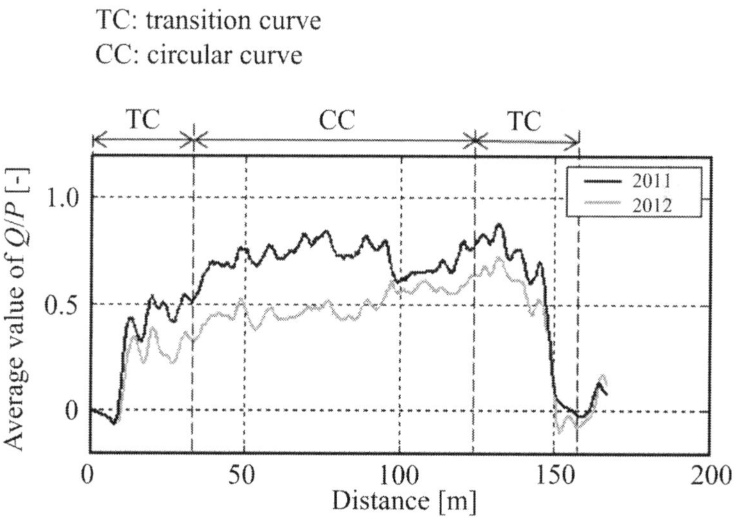

FIGURE 5.10 Comparison of derailment coefficients after averaging.

by making comparisons using the estimated torsional irregularity and continuous measurements by the PQ monitor bogie, it is possible to capture detailed changes in the derailment coefficient of the track.

5.1.4 Method for Estimating the Friction Coefficient between the Flange and Rail

When a railway vehicle passes through sharp curves, there is concern about the flange climbing over the outer rail. Generally, the running safety of the vehicle against flange climbing is evaluated by the value of Q / P. However, the ease of

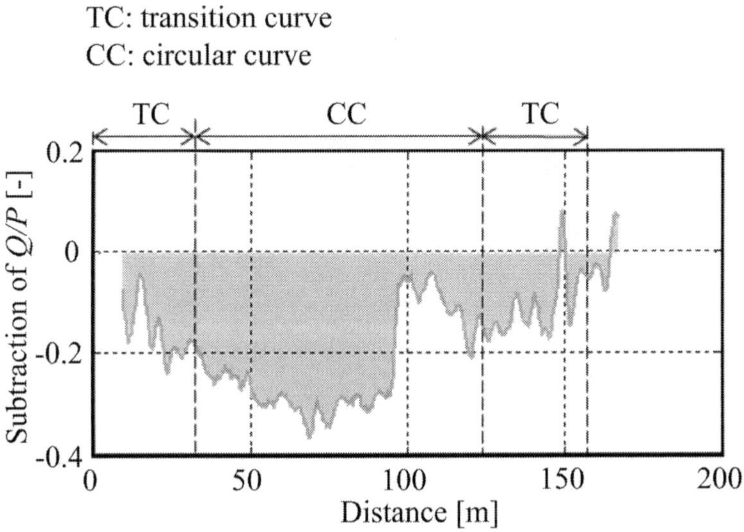

FIGURE 5.11 Temporal subtraction processing of derailment coefficients between two years.

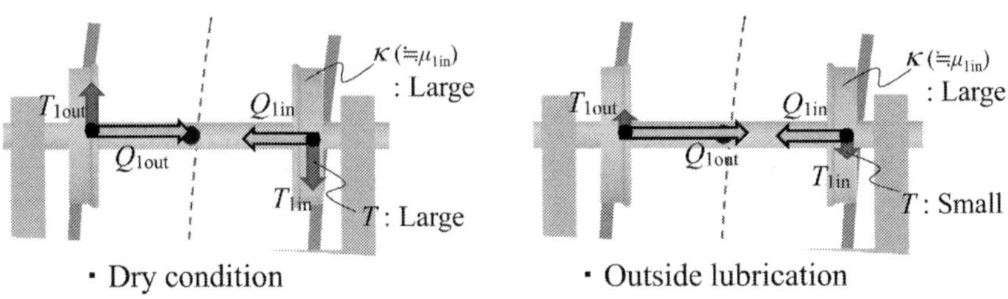

FIGURE 5.12 Relationship between the friction coefficient for the outside wheel and the steering moment.

flange climbing depends on the friction coefficient between the rail and leading outside wheel flange of the bogie (in this section, it is written as μ_{1out}). As described by Nadal's formula, the limit of derailment coefficient $(Q/P)_{CR}$ is calculated from μ_{1out} . The friction coefficient is likely to change dramatically over short spans due to the rail lubricators that supply grease to rail to prevent rail and wheel wear and the $(Q/P)_{CR}$ also changes. To evaluate the exact running safety of the vehicle quantitatively, μ_{1out} must be estimated [3].

Figure 5.12 illustrates the tangential force variation when the friction coefficient for the outer wheel changes. As can be seen from the figure, when the friction coefficient for the outer wheel becomes high, the tangential force increases. In view of this mechanism, we consider the method of estimating the friction coefficient

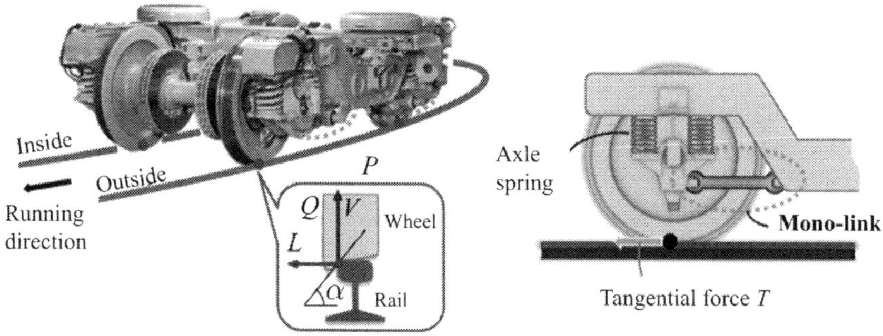

FIGURE 5.13 Bogie for monitoring wheel–rail contact forces.

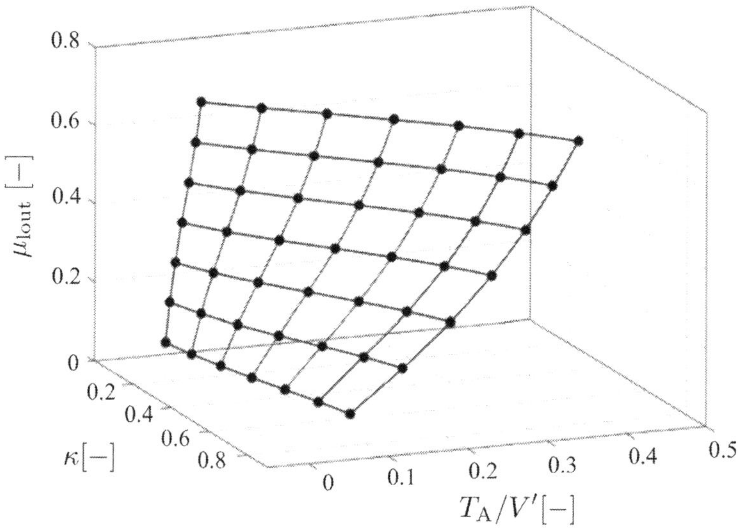

FIGURE 5.14 Lookup chart for estimating the friction coefficient $\mu_{1\mathrm{out}}$.

using the tangential force obtained by the PQ monitoring bogie's mono-link, as shown in Figure 5.13.

In this section, T_{X1} indicates the average value for the inside and outside wheels of the leading wheelset in the bogie. When $\mu_{1\mathrm{out}}$ is low, T_{X1} becomes small, and when $\mu_{1\mathrm{out}}$ is high, T_{X1} becomes large. To grasp the relationships between the friction coefficient and the contact forces, multibody dynamics simulations were carried out using SIMPACK, a software package extensively used in the railway field, and a vehicle model running on a sharp curve. $\mu_{1\mathrm{out}}$ and the friction coefficient for other wheels μ_{others} separately change from 0.1 to 0.7 in steps of 0.1, and Figure 5.14 shows the results under 49 conditions. The values are steady in the circular curve.

The graph shows the relationships between the estimated μ_{1out}, input κ and T_{X1} normalized by P_{1in}. The graph can be considered as a look-up table and can be defined for any curve radius. From the look-up table, it is possible to estimate the value of μ_{1out} from κ and T_{X1} / P_{1in}, that can be measured by the monitoring bogie. The limit of derailment coefficient $(Q / P)_{CR}$ also can be obtained from the estimated μ_{1out} based on Nadal's formula. Additionally, the Flange Climb Index (FCI) defined by dividing the measured Q / P by the estimated $(Q / P)_{CR}$ can be obtained. Figure 5.15 shows the overall flow of the estimation regarding running safety with a PQ monitoring bogie.

Figure 5.16 shows two examples of the derailment coefficient Q / P, κ and T_{X1} / P_{1in}. The average value of each is also shown. These were collected by a monitoring bogie on a sharp curve with a radius of 160 m.

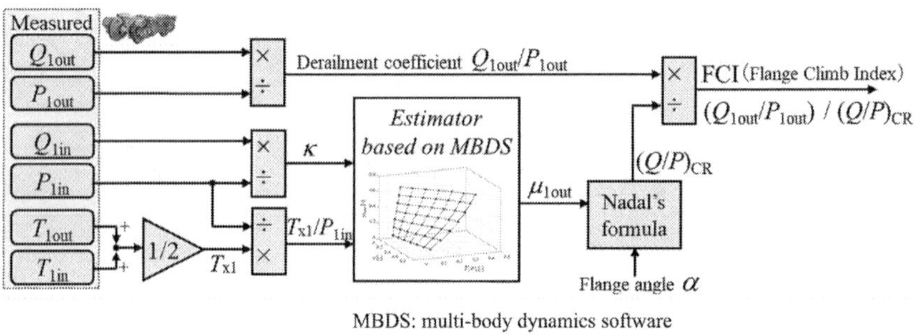

FIGURE 5.15 Flow of analysis of running safety against flange climbing.

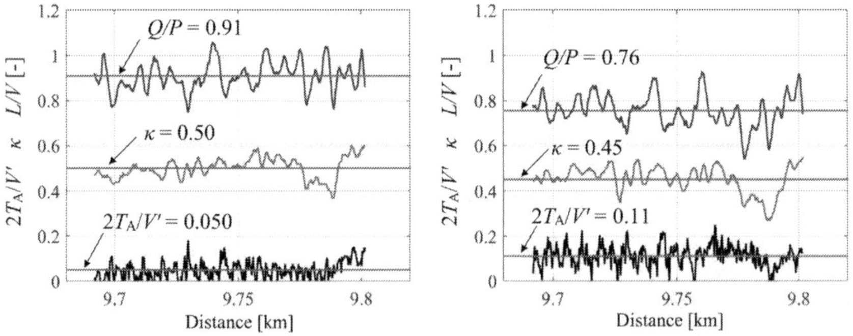

FIGURE 5.16 Examples of contact forces collected by a monitoring bogie.

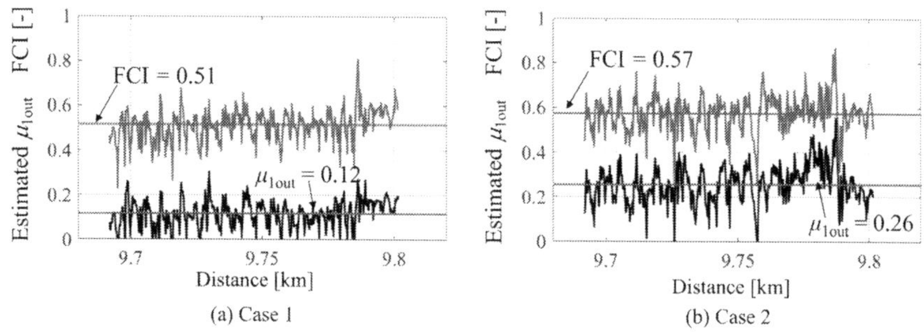

FIGURE 5.17 Examples of estimation of results μ_{1out} and the FCI.

Figure 5.17 shows μ_{1out} estimated by the method from the above inputs, and the FCI calculated from the measured Q/P and estimated $(Q/P)_{CR}$. In both examples, the value of κ is almost the same. This means that the value of the friction coefficient for the leading inner wheel is roughly the same in both cases. In case 2, the estimated μ_{1out} is higher than in case 1 because T_{X1}/P_{1in} is higher than in case 1. A high T_{X1}/P_{1in} value means that the steering force is high. A high steering force reduces the derailment coefficient, so in case 2, the derailment coefficient is small. Although the derailment coefficient is smaller in case 2, the value of the FCI is the same as in case 1 because $(Q/P)_{CR}$ is also low due to the low μ_{1out}. In the above results, because $(Q/P)_{CR}$ can be obtained by estimation of μ_{1out}, a more accurate analysis of the running safety of the vehicle against flange climb derailment, which could not be done with the derailment coefficient alone, can be made.

5.2 ESTIMATION OF TRACK IRREGULARITIES

5.2.1 Model-Based Approach

This section describes a method for estimating track longitudinal level irregularities, which causes a vibratory input to the car body, from the vertical car body accelerations using a KF; which is a typical state estimation method. To verify the validity of the estimation method, the results of estimating the track longitudinal level irregularity of a track using the measured vertical accelerations of the car body on a regional railway are shown [4].

5.2.1.1 Vehicle Model

In this study, a vehicle model representing the vertical motion of railway vehicles is used to construct the KF. The vehicle model is shown in Figure 5.18.

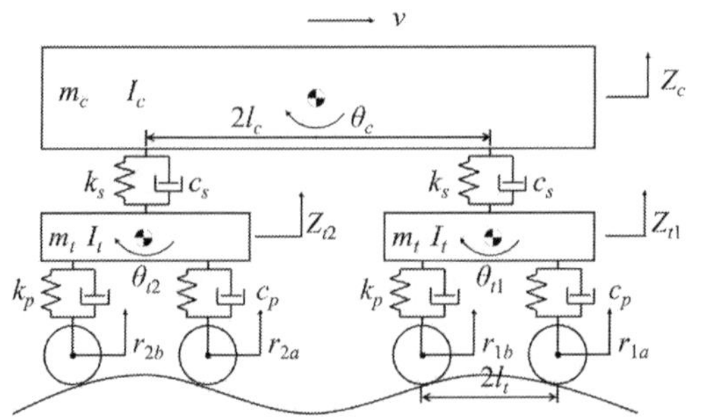

FIGURE 5.18 6-DOF vehicle model.

The model is a single-vehicle linear model that is assumed to run on a straight track, and it has six DOFs. There are two DOFs (vertical bounce and pitch) for the car body, two DOFs (vertical bounce and pitch) for the front bogie and two DOFs (vertical bounce and pitch) for the rear bogie (see section 2.4.1).

5.2.1.2 State Space Model

In track condition monitoring, the complexity and cost of the system would be reduced if the track condition could be determined solely from the motion of the railway car body. Therefore, a KF is constructed using only the information of the car body motion (in this case, bounce and pitching motion).

Equation (2.40) contains eight external force elements. Namely, the track displacements with respect to each of the four axles and their velocities. However, these eight inputs cannot be determined solely from the motion of the bogie. This is because information on the pitching motion of the bogie is not acquired.

The track geometry is expressed by the following state equation:

$$\underbrace{\begin{bmatrix} x(k) \\ x(k-1) \\ \vdots \\ \vdots \\ x(k-L+1) \end{bmatrix}}_{\boldsymbol{x}(k)} = \underbrace{\begin{bmatrix} 0 & \cdots & \cdots & \cdots & 0 \\ 1 & \ddots & \ddots & \ddots & \vdots \\ 0 & \ddots & \ddots & \ddots & \vdots \\ \vdots & \ddots & \ddots & \ddots & \vdots \\ 0 & \cdots & 0 & 1 & 0 \end{bmatrix}}_{F} \underbrace{\begin{bmatrix} x(k-1) \\ x(k-2) \\ \vdots \\ \vdots \\ x(k-L) \end{bmatrix}}_{\boldsymbol{x}(k-1)} + \underbrace{\begin{bmatrix} 1 \\ 0 \\ \vdots \\ \vdots \\ 0 \end{bmatrix}}_{L} w(k-1), \tag{5.2}$$

where $\boldsymbol{x}(k)$ is the track geometry and $w(k-1)$ is white Gaussian noise.

Equation (5.2) is simply a model of track displacements moving with time and does not include vehicle dynamics. Thus, the vehicle dynamics is expressed by the following measurement equation:

$$z(k)=\underbrace{\begin{bmatrix} h(0) & h(1) & \cdots & \cdots & h(L) \end{bmatrix}}_{H}\underbrace{\begin{bmatrix} x(k) \\ x(k-1) \\ \vdots \\ \vdots \\ x(k-L+1) \end{bmatrix}}_{x(k)}+v(k), \tag{5.3}$$

where $z(k)$ is the car body vertical acceleration and $v(k)$ is the sensor noise.

The vehicle model is expressed by the measurement equation as a convolution integral with an impulse response, as shown in Equation (5.3), where h in $\boldsymbol{H}$ is the unit impulse response of the car body vertical acceleration to the track geometry and L is the total number of such responses. The above state-space model allows the track geometry to be sequentially estimated from the vertical acceleration of the car body using the KF.

The impulse response of the vertical acceleration of the car body used in the measurement matrix varies with the travelling speed. In principle, the impulse response can be recalculated each time the travelling speed changes, but this is not practical in terms of computational load and efficiency.

Therefore, in this method, the impulse response is calculated in advance at each specified speed, and the impulse response corresponding to the speed at that time is used in the sequential estimation by the KF (Figure 5.19). In this study, the impulse responses were calculated using a vehicle model and used for numerical simulation.

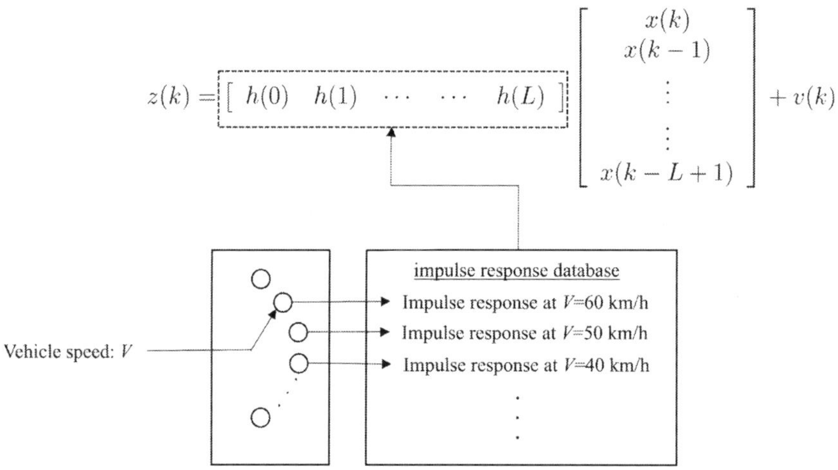

FIGURE 5.19 Selection of the impulse response for different speeds.

For linear systems such as the one treated here, the time response of the vehicle model can be calculated by using a finite impulse response filter. In other words, if the impulse response $h(\tau)$ of the vehicle model is known, the output (vertical acceleration of the car body) with respect to the input (track geometry) can be calculated by the following convolution integral and the vehicle model can be expressed by the measurement equation as in Equation (5.4):

$$z(t)=\int_0^t h(\tau)x(t-\tau)d\tau. \tag{5.4}$$

The unit impulse response of the vertical acceleration of the car body can be calculated by applying a unit impulse signal to the vehicle model. As an example, the impulse response for a travelling speed of 60 km/h, assuming a regional railway vehicle, is shown in Figure 5.20. The measurement point for the impulse response was just above the front bogie.

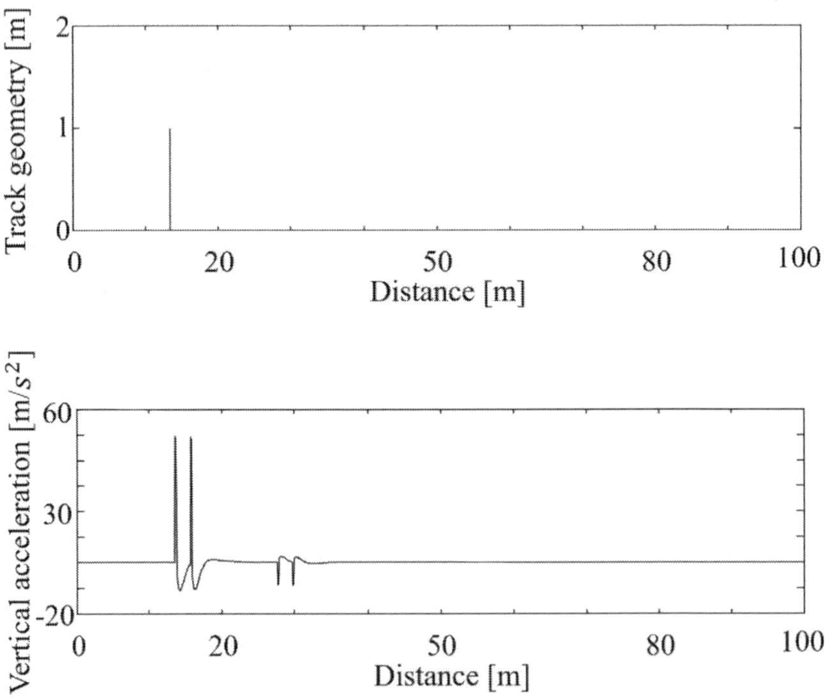

FIGURE 5.20 Impulse input and response (at $V = 60$ km/h).

5.2.1.3 Configuration of the Kalman Filter

The KF is used to construct an algorithm for estimating track geometry from the vertical acceleration of the car body.
The time update algorithm is

$$\hat{\boldsymbol{x}}(k|k-1)=\boldsymbol{F}\hat{\boldsymbol{x}}(k-1|k-1), \tag{5.5}$$

$$\boldsymbol{P}(k|k-1)=\boldsymbol{F}\boldsymbol{P}(k-1|k-1)\boldsymbol{F}^{T}+\boldsymbol{L}\boldsymbol{Q}\boldsymbol{L}^{T}. \tag{5.6}$$

The measurement update algorithm is

$$\boldsymbol{K}(k)=\boldsymbol{P}(k|k-1)\boldsymbol{H}^{T}\boldsymbol{S}(k|k-1)^{-1}, \tag{5.7}$$

$$\hat{\boldsymbol{x}}(k|k)=\hat{\boldsymbol{x}}(k|k-1)+\boldsymbol{K}(k)\tilde{z}(k|k-1), \tag{5.8}$$

$$\boldsymbol{P}(k|k)=\left[\boldsymbol{I}-\boldsymbol{K}(k)\boldsymbol{H}\right]\boldsymbol{P}(k|k-1), \tag{5.9}$$

where

$$\boldsymbol{S}(k|k-1)=\boldsymbol{H}\boldsymbol{P}(k|k-1)\boldsymbol{H}^{T}+\boldsymbol{R}, \tag{5.10}$$

$$\tilde{z}(k|k-1)=z(k)-\boldsymbol{H}\hat{\boldsymbol{x}}(k|k-1), \tag{5.11}$$

and the state transition matrix $\boldsymbol{F}$, the matrix $\boldsymbol{L}$ and the measurement matrix $\boldsymbol{H}$ are given in Equations (5.2) and (5.3). $\boldsymbol{P}$ is the error covariance matrix, $\boldsymbol{K}$ is the Kalman gain, $\boldsymbol{Q}$ is the covariance matrix for the system noise and $\boldsymbol{R}$ is the covariance matrix for the sensor noise.

5.2.1.4 Methods for Assessing Track Irregularities

Generally, track irregularities are measured and controlled by the 10-m chord versine method. In this method, a 10-m string is stretched over the rail and the separation between the rail and the string at its centre is measured (Figure 5.21) [4]. After the track geometry is estimated, it is then converted into longitudinal level irregularity by the 10-m chord versine method and evaluated.

The equation for calculating the longitudinal level irregularity from the track geometry is given below, where $a(x)$ is the longitudinal level irregularity from the 10-m chord versine method, $b(x)$ is the track geometry and x is the distance in metres.

$$a(x)=b(x)-\frac{b(x+5)+b(x-5)}{2}. \tag{5.12}$$

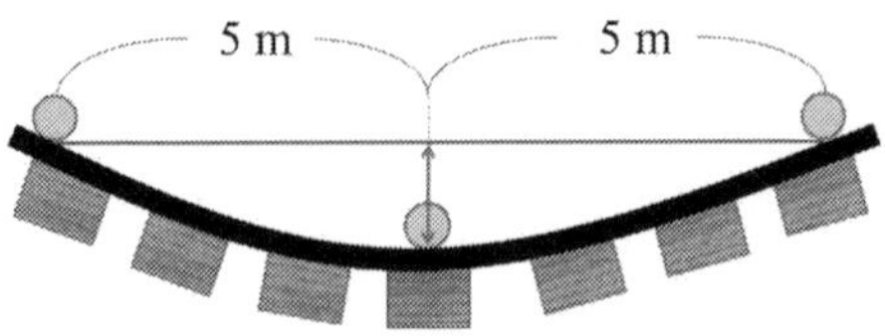

FIGURE 5.21 10-m chord versine method.

5.2.1.5 Evaluation of Estimation Methods by Simulation

To confirm the effectiveness of the track irregularity estimation method described above, the results of an estimation simulation on a regional railway are described. The simulation was carried out according to the following procedure:

- Track geometry data (wavelength > 6 m) simulating the spatial frequency characteristics of an actual track are created. Also, the speed data is set to simulate typical train operation.
- The track geometry and speed data are inputted into a vehicle model, and the vertical acceleration of the car body in the vertical direction just above the front bogie is calculated. Noise with a normal distribution is added to the vertical acceleration of the car body, which is then used as the measurement value.
- The track geometry is estimated from the measured vertical acceleration of the car body and the set velocity data using a KF.
- The track geometry and the estimated track geometry are each converted into longitudinal level irregularity by the 10-m chord versine method, and the estimation error in the 10-m chord versine method is determined. Evaluation is then carried out on the estimation accuracy of the results to confirm the validity of the method.

5.2.1.6 Generation of Vertical Acceleration of the Car Body

As no measured track geometry data was available, track geometry data (wavelength > 6 m) was generated from the spatial frequency characteristics of an actual track by generating random numbers. The speed profile was set assuming a single section of a typical regional railway.

The vertical acceleration of the car body just above the front bogie (calculated with track geometry and speed data input to the vehicle model) is shown in Figure 5.22. Table 5.1 shows the parameters of the vehicle model used in the calculations.

5.2.1.7 Assessment of Estimation Results and Estimation Accuracy

The estimated track geometry, the estimated longitudinal level irregularity using the 10-m chord versine method and the estimation error in the 10-m chord method are

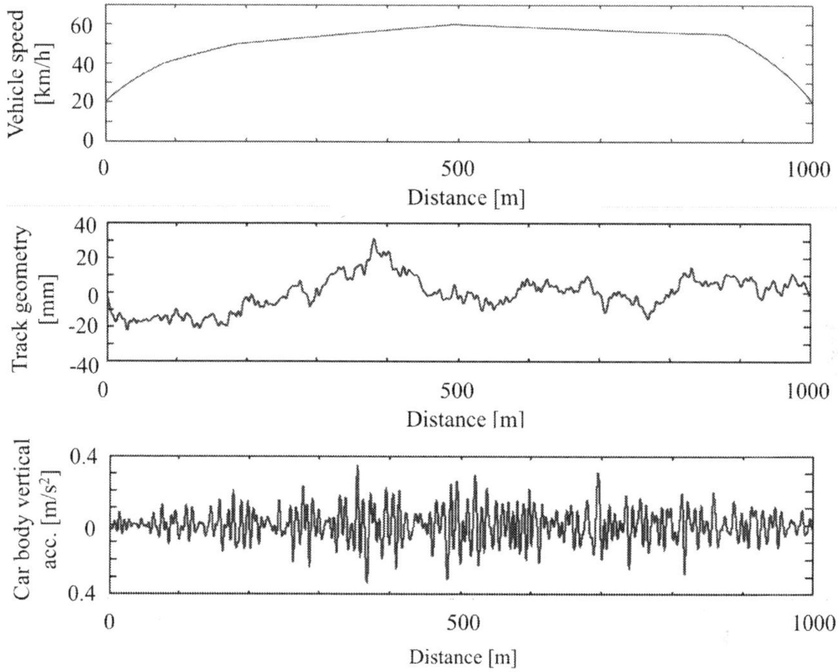

FIGURE 5.22 Vehicle speed, track geometry and car body vertical acceleration.

TABLE 5.1
Vehicle parameters

Symbol	Description	Unit	Value
m_c	Car body mass	kg	25,000
m_t	Bogie mass	kg	3100
I_c	Car body pitch inertia	kgm^2	856,940
I_t	Bogie pitch inertia	kgm^2	3417.8
$2l_c$	Car body base	m	14.1
$2l_t$	Wheel base	m	2.1
k_p	Primary suspension vertical stiffness	kN / m	2120
k_s	Secondary suspension vertical stiffness	kN / m	400
c_p	Primary suspension vertical damping	kNs / m	39.2
c_s	Secondary suspension vertical damping	kNs / m	96

shown in Figure 5.23. The variance of the system noise $w(k)$ and the measurement noise $v(k)$ in the estimation were set to $\sigma_w^2 = 1\times10^{-1}\,\mathrm{m}^2$ and $\sigma_v^2 = 5\times10^{-3}\,(\mathrm{m/s}^2)^2$, respectively.

From Figure 5.23, the estimated track geometry shows differences caused by the fact that the long-wavelength component cannot be estimated. This may be because the calculated vertical acceleration of the car body at the set travelling speed does not include a long-wavelength component for estimating the track geometry.

However, good accuracy estimation results were obtained for longitudinal level irregularities converted by the 10-m chord versine method. As the 10-m chord versine method is used for actual track maintenance and management in regional

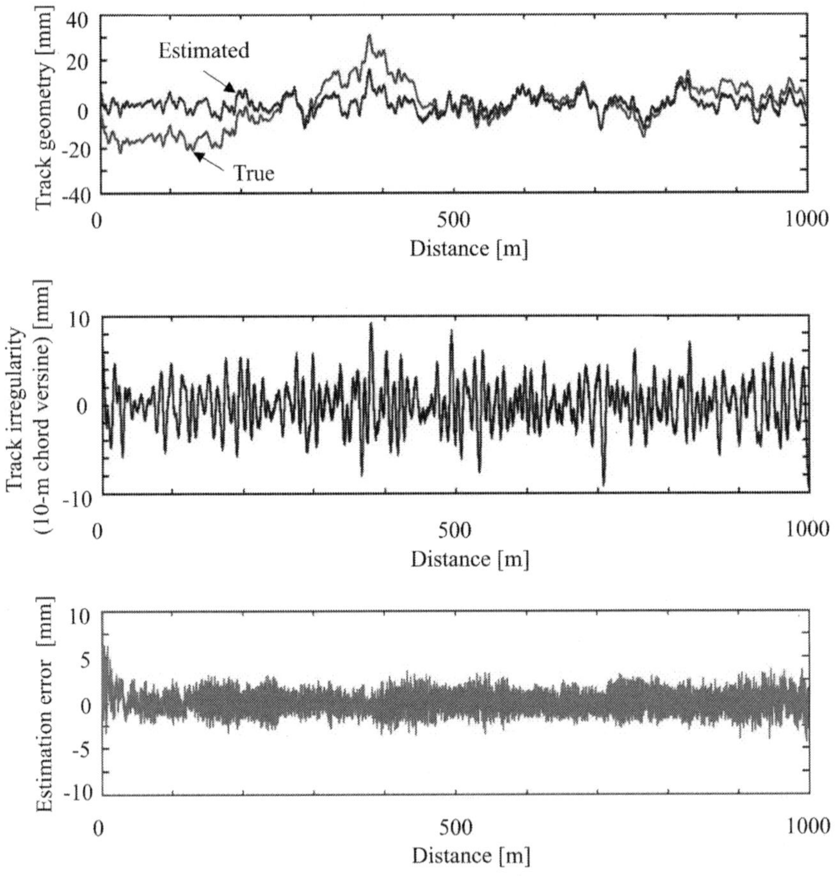

FIGURE 5.23 Estimation result of track geometry, estimated longitudinal-level irregularity and the estimation error in longitudinal-level irregularity.

TABLE 5.2
Evaluation result for the Sprague–Geers metric

Parameter	M	P	C
Value	0.01	0.03	0.03

railways (even if track geometry is not available) the estimation of the longitudinal level irregularity by this method is useful for track condition monitoring.

The Sprague–Geers metric [5], one of the Magnitude-Phase-Composite (MPC) metrics, was used to evaluate the estimation accuracy by comparing the estimated longitudinal level irregularity with the measured values. An MPC metric is a method for evaluating the correlation between two waveforms by focusing on the amplitude and phase of the waveforms. The equations for calculating each metric are shown below.

$$M = \sqrt{\frac{\sum e_i^2}{\sum m_i^2}} - 1 , \tag{5.13}$$

$$P = \frac{1}{\pi} \cos^{-1} \frac{\sum e_i m_i}{\sqrt{\sum e_i^2 \sum m_i^2}} , \tag{5.14}$$

$$C = \sqrt{M^2 + P^2} . \tag{5.15}$$

where M is the amplitude characteristic, P is the phase characteristic and C is the comprehensive evaluation index combining M and P. The closer each value is to 0, the higher the correlation with the measured value. In addition, e is the estimated value and m is the true (measured) value. The results of applying the Sprague–Geers metric to the estimated results of longitudinal level irregularity are shown in Table 5.2. Table 5.2 shows that each value is close to zero, indicating that there is a high correlation between the amplitude and phase.

5.2.1.8 Field Test

An on-board sensing device was installed next to the driver's cab of an in-service train, and a field test was conducted to measure car body vibration (Figure 5.24).

An onboard sensing device consisting of a three-axis accelerometer, a rate gyro and a GNSS receiver for identifying the vehicle's position was installed in the upper part of the front of the vehicle body to measure the vehicle vibration (Figure 5.24). The sampling frequency of the measured data was 82 Hz. This device was powered by the vehicle and continuously recorded the vehicle vibration and position during operation.

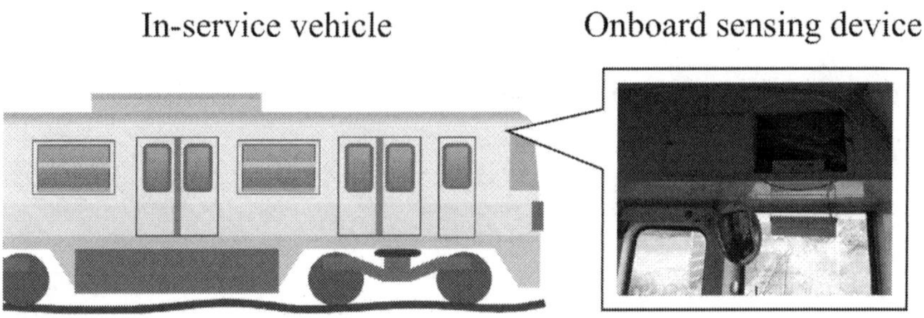

FIGURE 5.24 On-board sensing device.

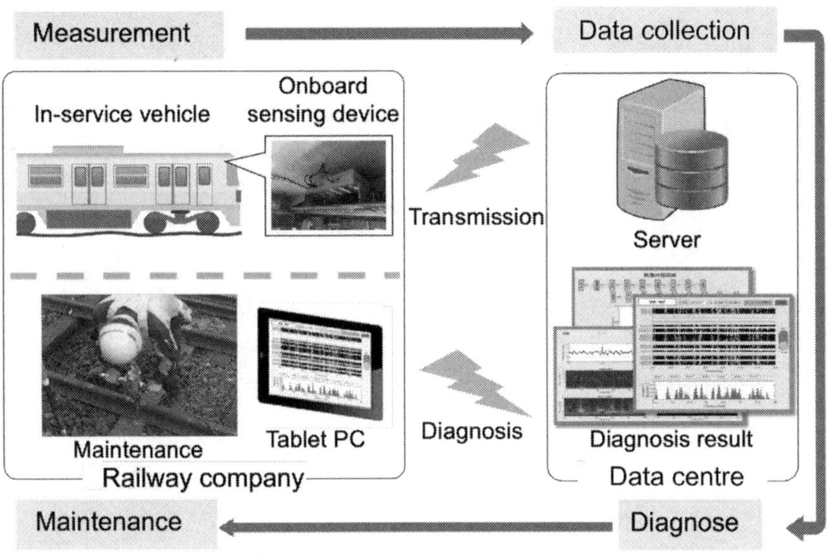

FIGURE 5.25 Track condition monitoring system.

The data used here was obtained by using the track condition monitoring system shown in Figure 5.25. The collected car body vibration data was transmitted to a server via a mobile phone network. A diagnosis based on the car body vibration data transmitted to the server was undertaken. Continuous monitoring of the track condition using this system enabled the detection of track faults.

The line for which the data were measured is a regional railway in the Tohoku region of Japan with the following characteristics: line length of 30.5 km, rail gauge of 1067 mm, 17 stations and non-electrified single-track type. The measurement dates were 5 February 2016, 20 May 2016, 24 June 2016 and 22 July 2016. The measurement items were vehicle position information, travelling speed, longitudinal acceleration, vertical acceleration, lateral acceleration and roll angular velocity. Although alignment and cross level irregularity can be evaluated from the measurement data, only longitudinal level irregularity was evaluated in this experiment.

5.2.1.9 Methods for Evaluating Track Conditions

The procedure for evaluating track conditions in this experiment is described below.

- Calculate the Root Mean Square (RMS) values from the measured vertical acceleration of the car body. From the RMS values of the vertical acceleration of the car body in all sections, extract the sections with the highest RMS values as the most significant for further attention.
- Use the KF to estimate the track geometry from the travelling speed and vertical acceleration of the car body in the extracted sections where the RMS values are significant. Convert the estimated track geometry to longitudinal level irregularity using the 10-m chord versine method.
- Compare the longitudinal level irregularity measured by the track inspection vehicle with the estimated longitudinal level irregularity to evaluate the accuracy of the estimation.

The measured travelling speed, vertical acceleration of the car body and RMS values calculated from the vertical acceleration of the car body are shown in Figure 5.26. Track section GH was selected because it includes the highest RMS values for the vertical acceleration of the car body among all sections.

5.2.1.10 Track Irregularity Estimation Results

Longitudinal level irregularities were estimated for the distances 13.7 km to 13.8 km from the start of the GH section using a KF. The data used for the estimation were from three runs on 20 May 2016 and 24 June 2016, for comparison with the track inspection vehicle data obtained on 30 May 2016. The travelling speeds and car body vertical acceleration used for the estimation are shown in Figure 5.27.

The estimated longitudinal level irregularities using the 10-m chord versine method are shown in Figure 5.28. The variance of the system noise w_n and the measurement noise v_n in the estimation was $\sigma_w^2 = 1\times10^{-1}\,\mathrm{m}^2$ and $\sigma_v^2 = 1\times10^{-2}\,(\mathrm{m/s}^2)^2$, respectively. The estimated longitudinal level irregularity is the average of the left and right rails.

From Figure 5.27, no significant differences were found in the measured travelling speed and vertical acceleration of the car body for the three runs, which suggests that the effect of the difference in travelling speed on the vertical acceleration of the car body is small. From Figure 5.28, no significant differences were also seen in the estimation results of the three runs. We can also expect that the estimation accuracy will be improved by averaging the results of multiple runs.

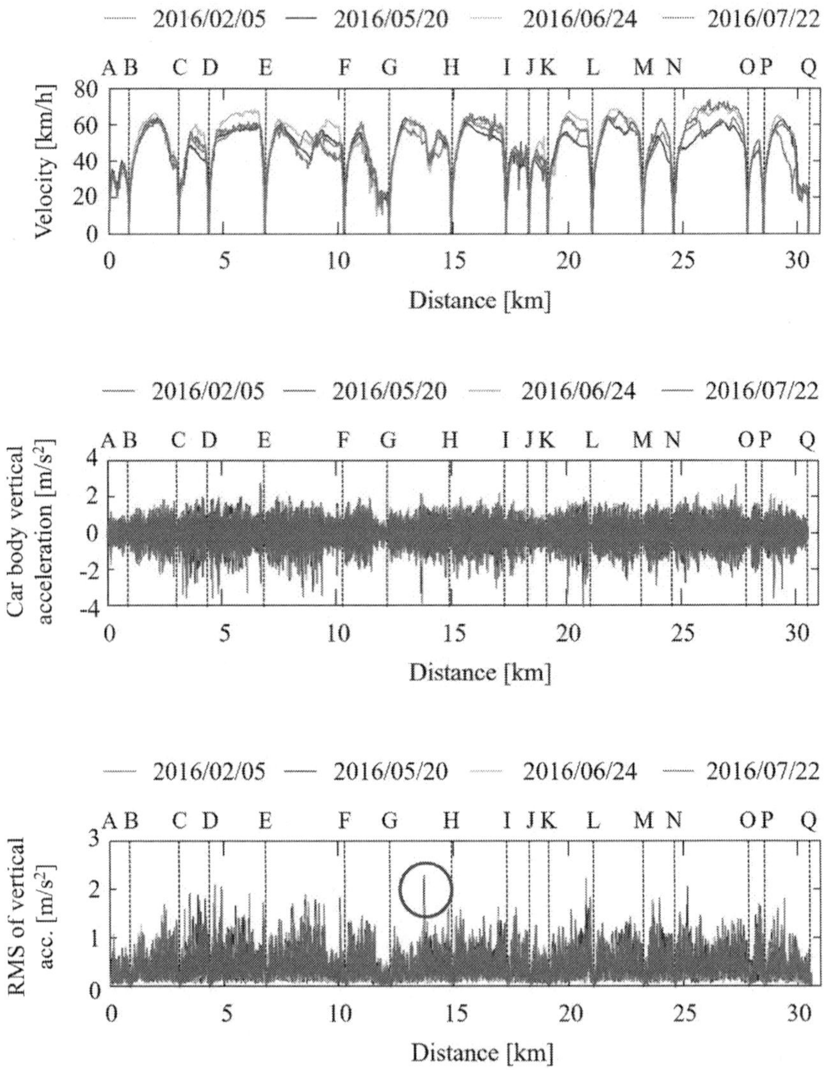

FIGURE 5.26 Measured car body vertical acceleration on a regional railway.

A comparison of the estimated longitudinal level irregularities and the longitudinal level irregularities measured by the track inspection car is shown in Figure 5.29. The estimation accuracy was evaluated using the Sprague–Geers metric.

The results shown in Table 5.3 show that the accuracy of the estimates for all three runs is almost the same, with significant longitudinal level irregularities of around 25 mm at 13.76 km.

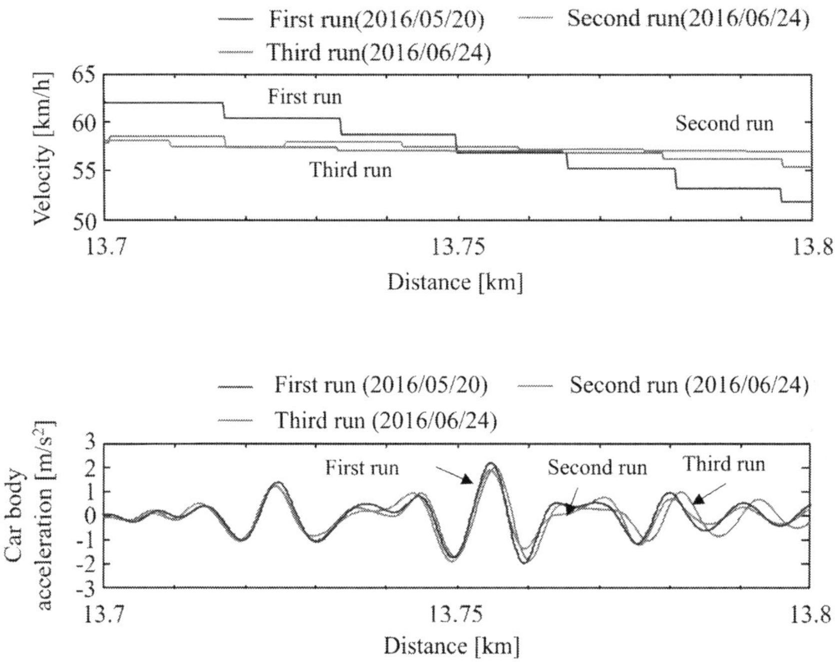

FIGURE 5.27 Measurement data for section GH.

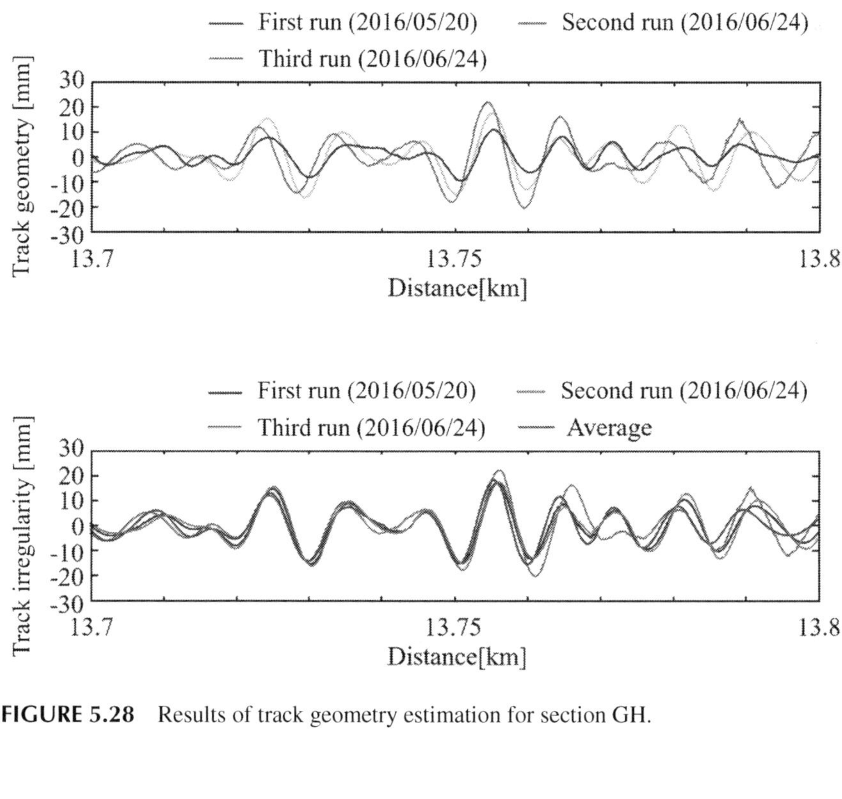

FIGURE 5.28 Results of track geometry estimation for section GH.

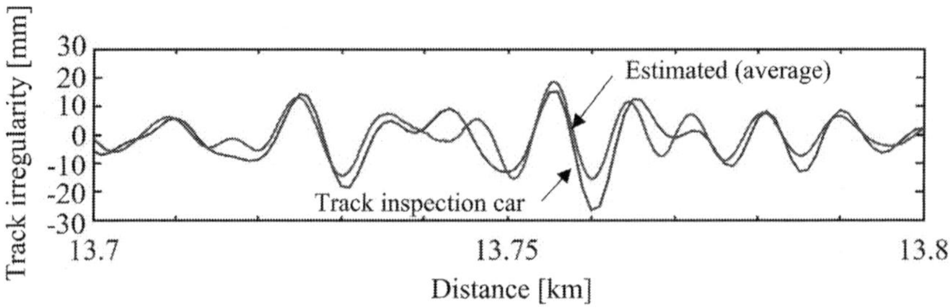

FIGURE 5.29 Comparison of estimated results with measurements obtained by track inspection vehicles on section GH.

TABLE 5.3
Evaluation of estimation results

Run	*M*	*P*	*C*
First run (2016/05/20)	-0.22	0.23	0.32
Second run (2016/06/24)	-0.13	0.21	0.25
Third run (2016/06/24)	-0.22	0.19	0.29
Average	-0.20	0.20	0.28

5.2.2 Data-Driven Approach

In this section, a data-driven approach is introduced for estimating track irregularities from measured car body vibration for track management [6]. The correlation between track irregularity and car body vibration was analysed using a multibody dynamics simulation of rail vehicles. Gaussian process regression [7] was applied to the track irregularity and car body vibration data obtained from the simulation, and a method was developed to estimate the track irregularities from the constructed regression model. The longitudinal level, alignment and cross level irregularities were estimated from the measured car body vibrations and travelling speeds on a regional railway, and the results were compared with the actual track irregularity data.

The track irregularity estimation procedure is shown in Figures 5.30 and 5.31 and given below:

- A railway vehicle model is prepared with multibody dynamics.
- From the track irregularity Power Spectral Density (PSD), track geometries for the profile, alignment and cross level are generated.
- The longitudinal level and alignment irregularities are calculated using the 10-m chord versine method.

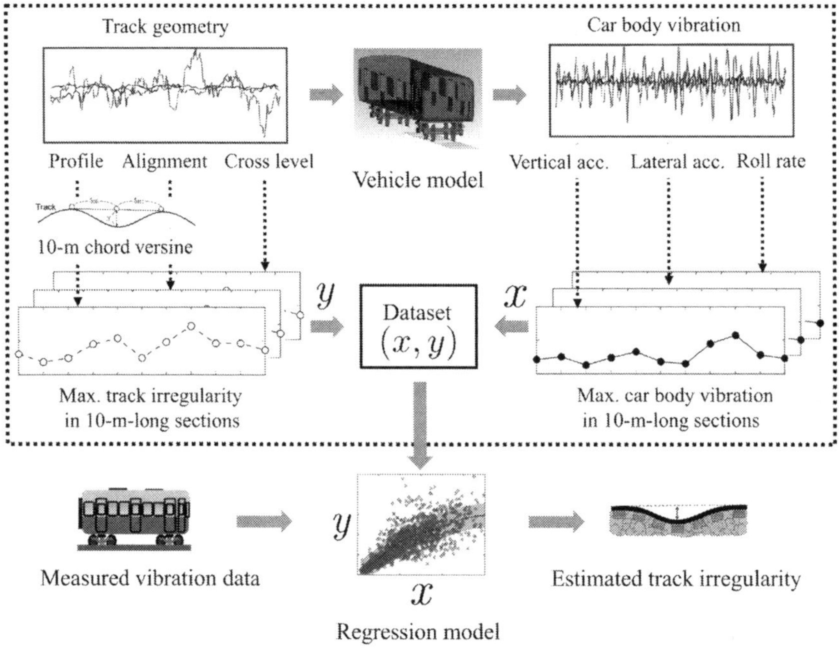

FIGURE 5.30 Track irregularity estimation procedure.

- The vehicle model is run on a track with the generated track geometries, and the vertical acceleration, lateral acceleration and roll rate are calculated.
- A dataset is created with the calculated maximum value of the car body vibration as input x and track irregularity as output y.
- GPR is applied to the dataset to create a regression model.
- The measured car body vibration of the actual vehicle is input into the constructed regression model to statistically estimate the track irregularity.

5.2.2.1 Generation of Track Irregularity and Car Body Vibration

5.2.2.1.1 Vehicle Model

The simulation model must output the vertical acceleration, lateral acceleration and roll rate of the car body with the track geometry, that is, profile, alignment and cross level, as track displacement. Therefore, a railway vehicle model with multibody dynamics is prepared, and a representative simulation is conducted. The dynamic simulations are per- formed using SIMPACK. Figure 5.32 shows the constructed vehicle model. The vehicle model comprises 7 rigid bodies (1 car body, 2 bogies and 4 wheelsets), each of which has 6 DOFs, resulting in a total of 42 DOFs. The car body, bogie and wheelset are connected and supported by spring and damper elements. By inputting the track geometry to the vehicle model and running the simulation, the vertical acceleration, lateral acceleration and roll rate of

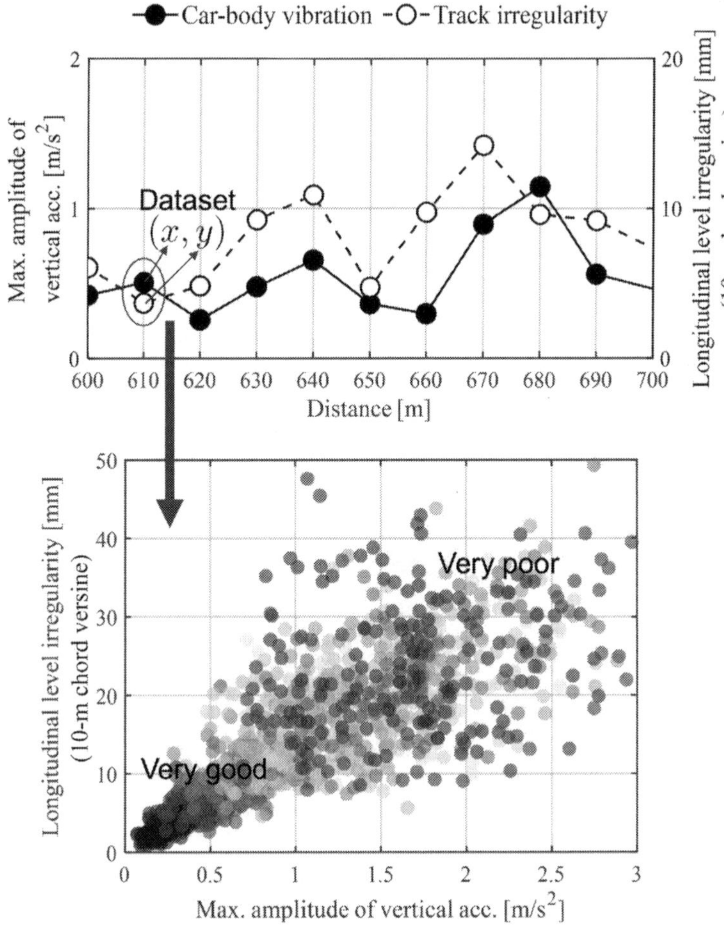

FIGURE 5.31 Dataset generation used for regression analysis.

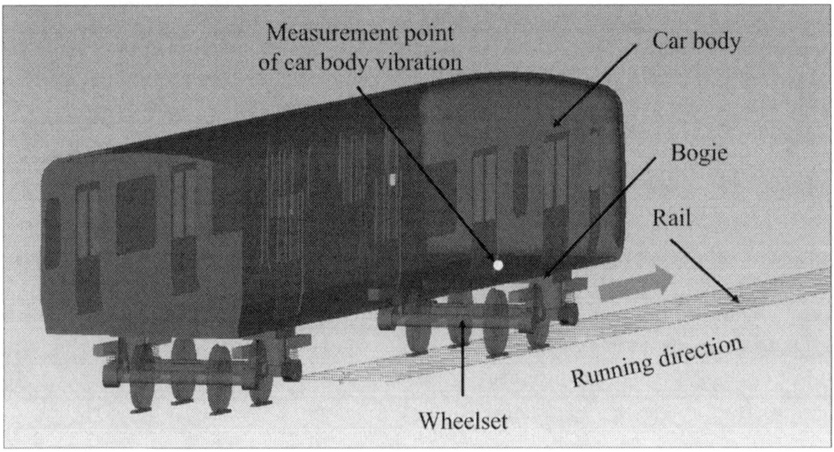

FIGURE 5.32 42-DOF vehicle model.

the vehicle, which are the vehicle motions just above the centre of the front bogie, are calculated.

The vehicle parameters used for the simulation are listed in Table 5.4, and they were set to match the car body vibration measured on a regional railway. The running distance was set to a straight section of 1000 m, and the sampling frequency was set to 100 Hz.

TABLE 5.4
Vehicle parameters

Description	Unit	Value
Car body mass	kg	25,000
Bogie mass	kg	3100
Wheelset mass	kg	1500
Car body inertia about x-axis	kgm^2	49,000
Car body inertia about y-axis	kgm^2	900,000
Car body inertia about z-axis	kgm^2	841,000
Bogie inertia about x-axis	kgm^2	2511
Bogie inertia about y-axis	kgm^2	1743.75
Bogie inertia about z-axis	kgm^2	1743.75
Wheelset inertia about x-axis	kgm^2	735
Wheelset inertia about y-axis	kgm^2	93.75
Wheelset inertia about z-axis	kgm^2	735
Car body base	m	14
Wheel base	m	2.1
Gauge	m	1.067
Wheel radius	m	0.43
Primary suspension vertical stiffness	kN/m	12,000
Secondary suspension vertical stiffness	kN/m	400
Primary suspension lateral stiffness	kN/m	6000
Secondary suspension lateral stiffness	kN/m	150
Primary suspension longitudinal stiffness	kN/m	8000
Secondary suspension longitudinal stiffness	kN/m	1000
Primary suspension vertical damping	kNs/m	40
Secondary suspension vertical damping	kNs/m	14
Primary suspension lateral damping	kNs/m	40
Secondary suspension lateral damping	kNs/m	180
Primary suspension longitudinal damping	kNs/m	40
Secondary suspension longitudinal damping	kNs/m	14

5.2.2.1.2 Track Model

The track PSD spectrum is widely used for dynamic simulations of railway vehicles [8, 9, 10]. The track PSD spectrum, characterised by a single-sided spectrum, is expressed as follows:

Profile:

$$S_p(\Omega) = \frac{A_p \Omega_c^2}{\left(\Omega^2 + \Omega_r^2\right)\left(\Omega^2 + \Omega_c^2\right)}. \tag{5.16}$$

Alignment:

$$S_a(\Omega) = \frac{A_a \Omega_c^2}{\left(\Omega^2 + \Omega_r^2\right)\left(\Omega^2 + \Omega_c^2\right)}. \tag{5.17}$$

Cross level:

$$S_c(\Omega) = \frac{\left(A_c \Omega_c^2 / a^2\right)\Omega^2}{\left(\Omega^2 + \Omega_r^2\right)\left(\Omega^2 + \Omega_c^2\right)\left(\Omega^2 + \Omega_s^2\right)}, \tag{5.18}$$

where $S_p(\Omega), S_a(\Omega)$ and $S_c(\Omega)$ are the PSDs of the track geometry (mm²/(rad/m)); Ω is the spatial angular frequency (rad/m); Ω_c, Ω_r and Ω_s are the critical spatial angular frequencies (rad/m); a is half of the nominal rolling circle distance of the wheel, and A_p, A_a and A_c are the roughness coefficients for the track geometry.

Parameters for the track PSD model are listed in Table 5.5.

TABLE 5.5
Track PSD model parameters and their interpretation

Parameter	Very Good	Baseline	Very Poor
A_p	1×10^{-5}	10×10^{-5}	80×10^{-5}
A_a	0.5×10^{-5}	5×10^{-5}	40×10^{-5}
A_c	0.1×10^{-6}	1×10^{-6}	35×10^{-6}
Ω_c (rad / m)	0.8	0.8	0.8
Ω_r (rad / m)	0.02	0.02	0.02
Ω_s (rad / m)	0.01	0.01	0.01

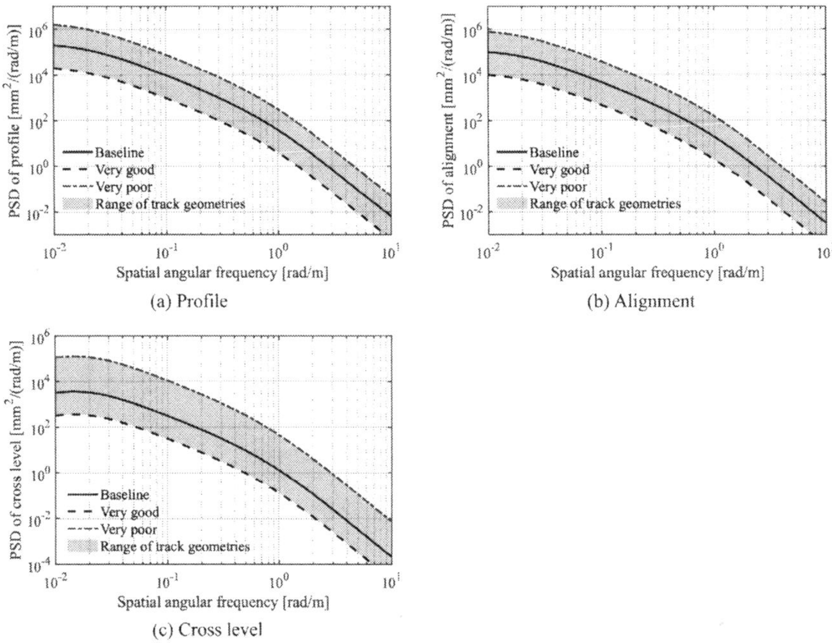

FIGURE 5.33 PSDs of the track geometry.

The roughness coefficients for the PSD functions for each track irregularity, A_p, A_a and A_c, are varied to generate 12 tracks from good to degraded conditions. The longitudinal level and cross level roughness coefficients, A_p and A_c, respectively, are usually the same, but they are listed separately to allow for independent assessment of track irregularity types. As a representative example, the PSDs for the best, nominal and worst conditions of the profile, alignment and cross level of the track geometry are shown in Figure 5.33, and the track geometry generated by the PSDs is shown in Figure 5.34.

The track geometry shown in Figure 5.34 was converted to track irregularities using the 10-m chord versine method by applying Equation (5.12), and regression analysis was performed.

5.2.2.2 Regression Analysis of Car Body Vibration and Track Irregularity

5.2.2.2.1 Gaussian Process Regression

Here the relationship between the car body vibration and track irregularity is examined using GPR, which is a nonlinear regression method. GPR is a nonparametric Bayesian approach to nonlinear regression problems that can provide uncertainty in predictions. GPR defines two elements: training data comprising

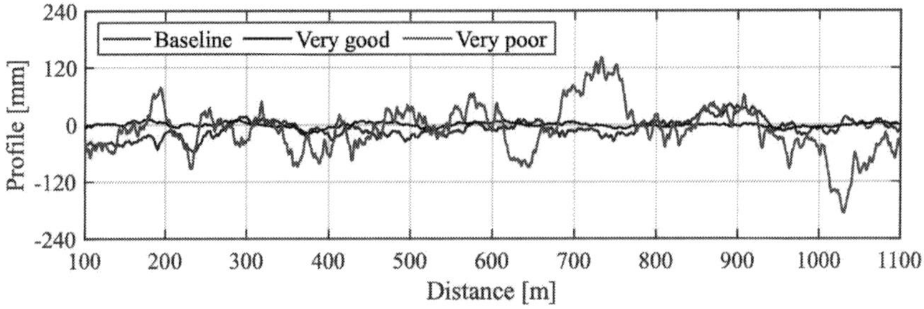

(a) Profile

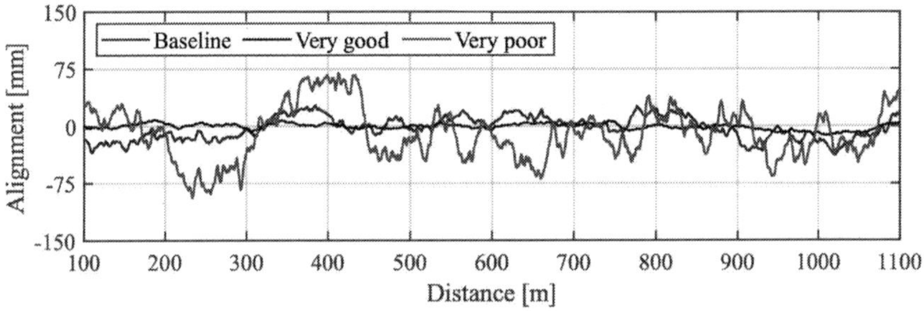

(b) Alignment

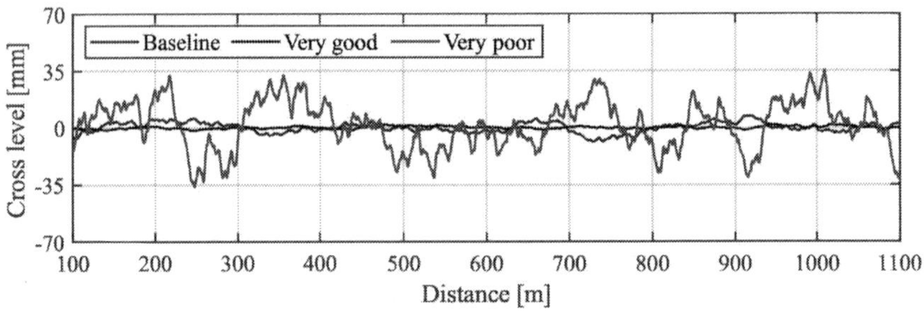

(c) Cross level

FIGURE 5.34 Generated track geometry.

pairs of input and output y and x, kernel function $k(x, x')$ providing the covariance of the Gaussian distribution, which is the similarity between inputs x and x'.

Given the training data $\mathcal{D} = \{(x_1, y_1), \cdots, (x_N, y_N)\}$, the output of the test data for a new input are expressed as follows:

$$p(y^*, x^*, \mathcal{D}) = N\left(\boldsymbol{k}_*^T \boldsymbol{K}^{-1} \boldsymbol{y}, k_{**} - \boldsymbol{k}_*^T \boldsymbol{K}^{-1} \boldsymbol{k}_*\right), \tag{5.19}$$

where

$$\begin{aligned}
\boldsymbol{K} &= \begin{bmatrix} k(x_1,x_1), & k(x_1,x_2) & \cdots & k(x_1,x_N) \\ k(x_2,x_1), & k(x_2,x_2) & \cdots & k(x_2,x_N) \\ \vdots & \vdots & \ddots & \vdots \\ k(x_N,x_1), & k(x_N,x_2) & \cdots & k(x_N,x_N) \end{bmatrix}, \\
\boldsymbol{k}_* &= \left[k(x^*,x_1),k(x^*,x_2),\cdots,k(x^*,x_N)\right]^T, \\
k_{**} &= k(x^*,x^*), \\
\boldsymbol{y} &= \left[y_1,y_2,\cdots,y_N\right]^T.
\end{aligned}$$

A radial basis function kernel and white kernel is used, described as:

$$k(x,x') = \theta_1 \exp\left(-\frac{|x-x'|^2}{\theta_2}\right) + \theta_3\delta(x,x'), \tag{5.20}$$

where θ_1, θ_2 and θ_3 are hyper-parameters and δ is the Kronecker delta, which is 1 when $x = x'$ and 0 when $x \neq x'$.

The white kernel [second term in Equation (5.20)] is a kernel function on the magnitude of noise in the objective variable and is useful when the dataset contains noise. In Equation (5.20), the hyper-parameters $(\theta_1, \theta_2$ and $\theta_3)$ control the extent to which the data are similar. Optimising the hyper-parameters therefore improves the performance of fitting to the measured data. The hyper-parameters are optimised by the maximum-likelihood estimation method.

$$E\left[y^*,x^*,\mathcal{D}\right] = \boldsymbol{k}_*^T \boldsymbol{K}^{-1}\boldsymbol{y}, \tag{5.21}$$

$$V\left[y^*,x^*,\mathcal{D}\right] = k_{**} - \boldsymbol{k}_*^T \boldsymbol{K}^{-1}\boldsymbol{k}_*. \tag{5.22}$$

Figure 5.35 shows the GPR results for the vertical acceleration and longitudinal level irregularity, lateral acceleration and alignment irregularity, and roll rate and cross level irregularity. The results of the linear regression without an intercept are also shown for comparison. Figure 5.35 shows that the characteristics are different between regions with a small acceleration and roll rate (good track condition) and those with a large acceleration and roll rate (poor track condition). The variance in the data for the longitudinal level, alignment and cross level irregularities is large for the poor track condition. Nonlinearity appears in the GPR regression curve in these regions, and the difference from the regression line suggests that GPR is a more effective regression method. To adapt to the actual line, it is necessary to make estimates using the measurement data. Therefore, it is necessary to generate

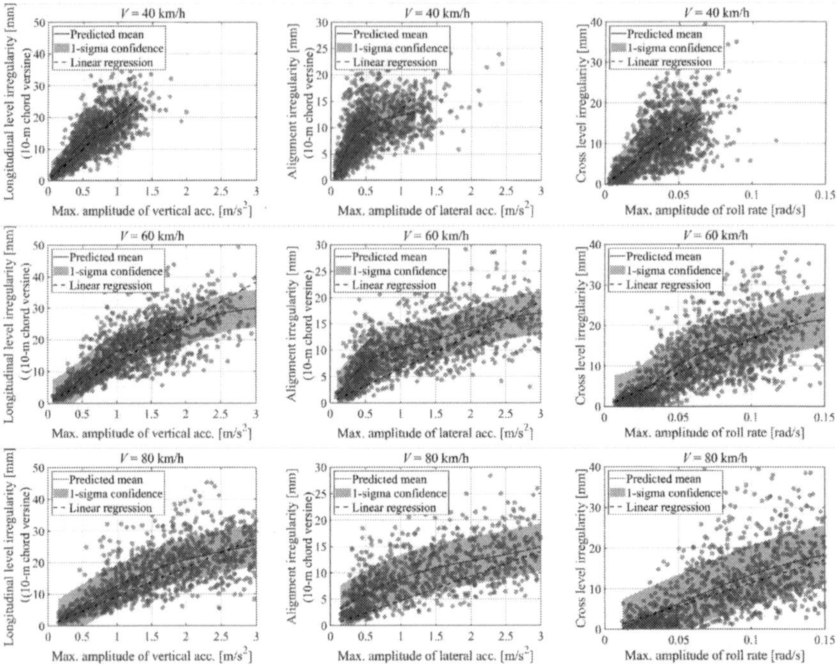

(a) Longitudinal level (b) Alignment (c) Cross level

FIGURE 5.35 Results of GPR.

a new dataset by adding the measurement data to the dataset generated by the simulation and then make the estimates.

5.2.2.3 Application to Track Condition Monitoring in a Regional Railway

Figure 5.36 shows the method for estimating track irregularities using GPR. First, the correlation between the car body vibration and track irregularities corresponding to the travelling speed is calculated in advance using GPR. Next, the maximum amplitude values of the vertical acceleration, lateral acceleration and roll rate measured by the track condition monitoring system for each 10-m-long section and the average travelling speed are calculated and used as inputs to the estimator. The estimator outputs longitudinal level, alignment and cross level irregularities corresponding to the travelling speed along with a 1σ confidence region.

Figure 5.37 indicates that most of the longitudinal level, alignment and cross level irregularities estimated using the GPR method are within the 1σ confidence region, confirming the effectiveness of this method. However, in some sections, the

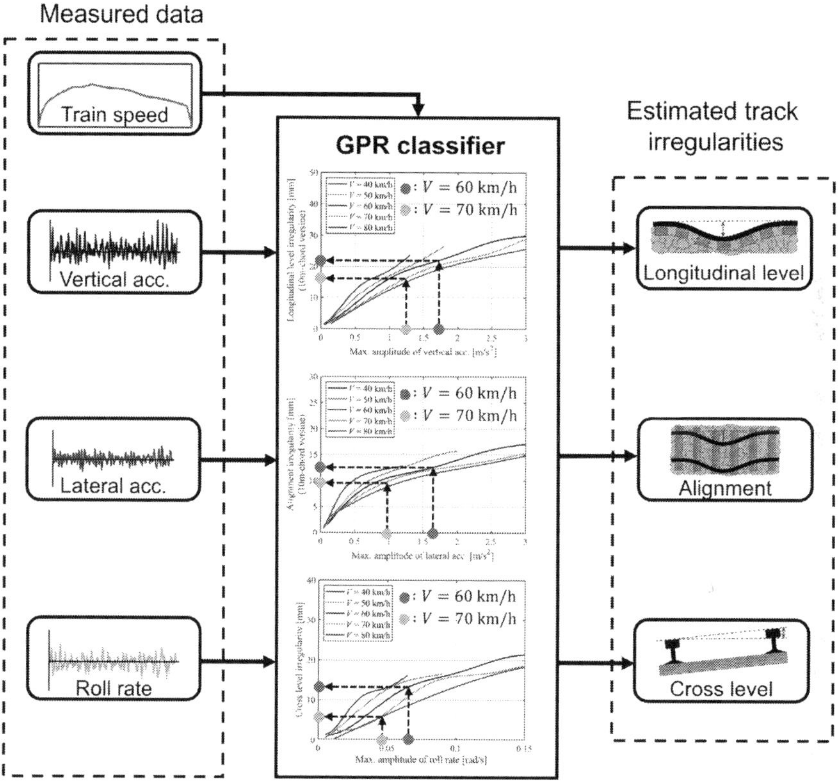

FIGURE 5.36 Track irregularity estimation procedure using a GPR classifier.

errors between the estimated results and measured data are large. This may be due to factors other than the track irregularities.

Comparison of the estimation results from GPR with those from linear regression shows no significant differences for longitudinal level irregularity and cross level irregularity. Conversely, large differences can be observed in alignment irregularities. This can be understood from the fact that there are large differences between linear regression and GPR for alignment irregularities. GPR also has the advantage that, unlike linear regression, it can present confidence regions at the same time.

In Figure 5.37, there are points where the estimated and measured values differ significantly, but such points may be affected by factors other than track irregularities. Therefore, it is considered that further useful information for maintenance could be obtained by investigating these locations.

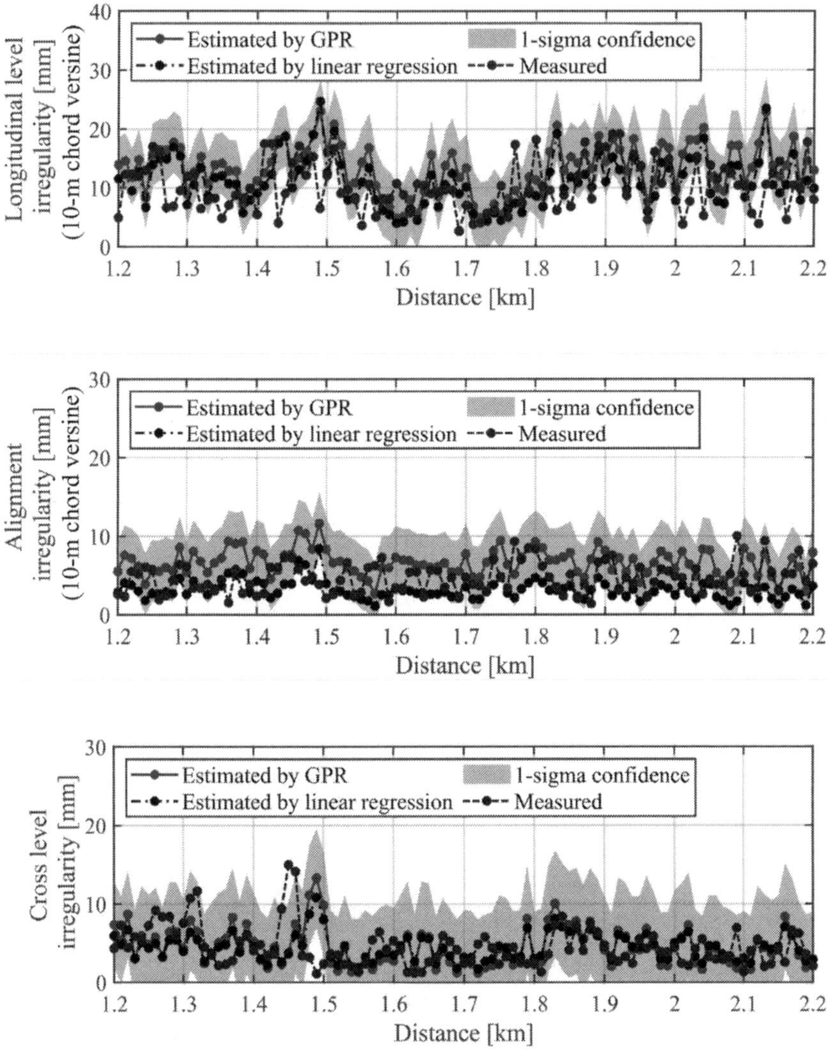

FIGURE 5.37 Estimated track irregularities in a section between 1.2 km and 2.2 km.

The data-driven approach presented here is an estimation method that is simpler than analytical or numerical models, and so it has the potential to be applied to many regional railways. However, the estimation accuracy depends on the amount of data collected and the regression method used, so further investigation of the estimation accuracy is needed. This chapter presents a basic investigation of the applicability of the data-driven approach, and in the future, it will be necessary to investigate the amount of data collected and the regression method used from the perspective of estimation accuracy.

5.3 LOW-ADHESION DETECTION IN THE WHEEL–RAIL INTERFACE

5.3.1 Low Adhesion Background and Physics

Low adhesion in railway systems is a significant issue often misunderstood by industry and general public alike. There is currently a lack of robust information about the changing picture of areas of low adhesion with respect to short term trends (over a daily period) and macro trends (across the seasons). A full understanding of the contemporary railhead condition across a network is currently unfeasible with the level of technology deployed.

The efficiency of rail travel is principally due to the minimal deformation and therefore small contact patches formed between a steel wheel on a steel rail rolling contact, but this fundamental property exposes risks that relatively minor contaminants within this area can have significant consequences. Low adhesion primarily arises due to this phenomenon, with contaminants forming a so-called 'third body layer'. These can be made of:

- Leaves and organic matter: During autumn, falling leaves can be crushed under the wheels, forming a slick, slippery substance.
- Moisture, condensation and 'micro-wetting': Rain, dew, or condensation can create a thin water film, further exacerbated by oil or grease.
- Contaminants from railway operations: Residues from braking systems, such as oil, and iron oxides, contribute to a reduction in friction.

The effects of low adhesion manifest as wheel slip events in traction or, perhaps more significantly, wheel slide in braking that can lead to increased operational risks such as station stops being missed or SPAD (Signals Passed At Danger) events. Beyond traction and braking, low adhesion impacts the guidance dynamics of the vehicle as the creep forces essential for stability are affected by the changing coefficient of friction. This can lead to increased wear and tear on both the wheel and the railhead resulting in more frequent maintenance and higher operational costs. Typical mitigation activity is often the instruction to undertake defensive driving techniques over large (and potentially unaffected) parts of the network, that in turn cause disruption to timetabling. In the UK, it has been estimated that low adhesion events can result in annual costs exceeding £350 million due to delays, damage to equipment, and increased maintenance requirements such as rail head cleaning. Fundamental methods for identifying adhesion events from wheel slip (under traction) or wheel slide (under braking) exist [11, 12] but methods for monitoring low adhesion across a rail network is an outstanding research challenge for the rail industry.

5.3.2 Model-Based Approaches to Low Adhesion Estimation

A change in the coefficient of friction directly results in a change in the interaction mechanisms between the wheel and the rail. This manifests as changes in both the traction and braking dynamic, but also in the guidance and stability aspects of vehicle motion. The principle behind adhesion estimation in this section is to measure the inertial properties of the moving parts of the running gear in the key guidance modes of motion, and process this information to attribute changing properties to adhesion changes. This section explores the use of a Kalman filter implementation that can estimate the contact force in the wheel/rail interface using a linear suspension model in the filter formulation.

The Kalman-Bucy is a mature filtering methodology that has been used for a wide variety of applications. There are discrete time [13] and continuous time [14] variants of this algorithm. The former will be how the method will be applied in the long term, but the latter is demonstrated here due to the continuous nature of the system dynamics. The fundamental state space equations of the system dynamics can be defined as

$$\dot{x} = Ax + Bu + B_w \tilde{w} , \tag{5.23}$$

$$z = Cx + Du + D_w \tilde{w} + \tilde{v} , \tag{5.24}$$

where $\boldsymbol{u}$ are the known inputs, $\tilde{\boldsymbol{w}}$ is the process noise, and $\tilde{\boldsymbol{v}}$ is the measurement noise. The minimisation of the steady-state error covariance

$$P = \lim_{t\to\infty} E\left[\{ x - \hat{x} \}\{ x - \hat{x} \}^T\right]. \tag{5.25}$$

The solution is of the form

$$\dot{\hat{x}} = F\hat{x} + Gu + K\left(z - C\hat{x} - Du\right) . \tag{5.26}$$

The Kalman filter gain $\boldsymbol{K}$ is determined through a Riccati equation to be

$$K = \left(PC^T + \bar{N}\right)\bar{R}^{-1} , \tag{5.27}$$

where

$$\bar{R} = R + HN + N^T H^T + HQH^T , \tag{5.28}$$

$$\bar{N} = G\left(QH^T + N\right) , \tag{5.29}$$

and $E\left[\tilde{w}\right] = E\left[V\right] = 0, E\left[\tilde{w}\tilde{w}^T\right] = Q,\ E\left[\tilde{v}\tilde{v}^T\right] = R,\ E\left[\tilde{w}\tilde{v}^T\right] = N$

The contact force is estimated included by augmenting the state matrix to include 'Force' as an unmeasured variable which acts on the system states. The model-based filter therefore requires:

- An accurate linearised suspension model. The dynamic equations that represent the guidance motion of the vehicle in terms of suspension parameters is characterised into a state-space and populates the Matrices $\boldsymbol{A}, \boldsymbol{B}, \boldsymbol{C}$ and $\boldsymbol{D}$
- The tuning of the matrices $\boldsymbol{Q}$ and $\boldsymbol{R}$. The former represents the user's assumption on the certainty of the state equation, with the later the assumption on the certainty of the measurement. They represent a trade-off where only one can be minimised at a time, namely, you either trust the model and the measurement equally, or one more than the other.

The guidance dynamics of a conical wheelset can be thought of as two interacting principles: the yaw dynamics and the lateral dynamics. The basic principle of which being that a yaw angle will cause the wheelset to move laterally, and that a lateral movement will create a yaw movement due to the increasing conicity. An excellent explanation of these dynamics can be found in [15].

This principal motion can be expressed by these two simplified, coupled equations:

$$m_w \ddot{y}_w = F_{susp} + F_{cont}, \tag{5.30}$$

$$J_w \ddot{\varphi}_w = M_{susp} + M_{cont}, \tag{5.31}$$

where m_w is the mass of the wheelset, y_w is the lateral movement of the wheelset, F_{susp} are the lateral suspension forces, F_{cont} are the wheel/rail contact forces, J_w is the yaw moment of inertia, φ_w is the yaw angle of the wheelset, M_{susp} is the moments created by the suspension in yaw and M_{cont} is the contact forces related into yaw moments.

The 'knowns' in the equations are related to the suspension characteristics and the masses. The 'unknowns' are the contact conditions. The Kalman filter can be formed in such a way to estimate these 'unknowns' by using a linear model of the suspension systems to provide an insight to the numerical quantities of F_{susp}, and resultant estimation of F_{cont} as an unmeasured state provides an analogy of adhesion conditions.

There are two key approaches to generating a linear model of the suspension arrangement: through first principles analysis whereby the suspension system is expressed as interacting dynamic equations representing the geometry and interacting forces; or via systems identification techniques where the dynamic equations are attributed automatically by data analysis. Both methods are equally applicable if a validated linear model is obtained.

The formulation of the linear model here is based on a plan-view half vehicle model. This can be thought of as lumped masses representing the wheelsets, bogie and vehicle body connected via stiffness and damping components at certain physical positions between the bodies, and constrained to move only in the horizontal plane. Analysis of this geometric arrangement leads to linear equations of motion describing the suspension movement due to the contact forces acting on the wheelsets.

It is helpful in this case study to arrange these into a state space description of the form:

$$\dot{\boldsymbol{x}} = \boldsymbol{A}\boldsymbol{x} + \boldsymbol{B}_w \tilde{\boldsymbol{w}}, \tag{5.32}$$

$$z = \boldsymbol{C}\boldsymbol{x}, \tag{5.33}$$

where $\boldsymbol{x}$ is a vector that contains the system states (principally the position and velocity of the lumped masses in the lateral and yaw motions). $\boldsymbol{A}$ is a square matrix that contains a description of the stiffness and damping components within the system, formed from the Newton/Euler analysis of the suspension. $\tilde{\boldsymbol{w}}$ contains a vector of the track disturbances which is mapped to the states via the matrix $\boldsymbol{B}_w$, essentially representing the forces acting on the wheelset. The suspension system has no controlled input so the usual $\boldsymbol{Bu}$ term is omitted. z is a vector that contains the measured variables of the system and $\boldsymbol{C}$ maps the states to these measurements.

The choice of states $\boldsymbol{x}$ such as to ease the mapping of state to measurement and are formed in the following way. Consider the primary suspension connection of the bogie frame to the wheelset. The forces generated across this suspension are a result of the differential position and velocity between the two bodies as a result of compression/extension of springs and the excitation of dampers. The movement of the primary suspension can be described by the variables within the vector x_{ff} (ignoring longitudinal modes):

$$x_{ff} = \begin{bmatrix} y_{FB} - y_{FF} & \dot{y}_{FB} - \dot{y}_{FF} & \psi_{FB} - \psi_{FF} & \dot{\psi}_{FB} - \dot{\psi}_{FF} \end{bmatrix}^T, \tag{5.34}$$

where y represents the absolute lateral position of the leading (front) bogie (subscript FB) and the leading (front) wheelset (subscript FF), and ψ is the absolute angle in radians. It is assumed both y and ψ are measured from a shared frame of reference such as the track centre line.

This expression can be simplified by replacing terms such as $y_{FB} - y_{FF}$ with equivalent term $y_{\delta FF}$ to provide the form:

$$\boldsymbol{x}_{ff} = \begin{bmatrix} y_{\delta FF} & \dot{y}_{\delta FF} & \psi_{\delta FF} & \dot{\psi}_{\delta FF} \end{bmatrix}^T. \tag{5.35}$$

Similarly for the trailing (rear) wheelset:

$$\boldsymbol{x}_{fr} = \begin{bmatrix} y_{\delta FR} & \dot{y}_{\delta FR} & \psi_{\delta FR} & \dot{\psi}_{\delta FR} \end{bmatrix}^T. \tag{5.36}$$

And for the leading bogie:

$$\boldsymbol{x}_{fb} = \begin{bmatrix} y_{\delta FB} & \dot{y}_{\delta FB} & \psi_{\delta FB} & \dot{\psi}_{\delta FB} \end{bmatrix}^T, \tag{5.37}$$

where both $y_{\delta FB}$ and $\psi_{\delta FB}$ are the differential position of the bogie and the vehicle.

The states of the half vehicle model are therefore:

$$x = \begin{bmatrix} \boldsymbol{x}_{ff} \\ \boldsymbol{x}_{fr} \\ \boldsymbol{x}_{fb} \end{bmatrix}. \tag{5.38}$$

The vector $\tilde{\boldsymbol{w}}$ comprises of the unmeasured disturbances into the system, which in this case will be considered as the contact forces acting on the wheelset. This can be defined as:

$$\tilde{\boldsymbol{w}} = \begin{bmatrix} F_{yFF} & M_{zFF} & F_{yFR} & M_{zFR} \end{bmatrix}^T, \tag{5.39}$$

where F_y is the lateral contact force acting on the leading (front) or trailing (rear) wheelset of the leading (front) bogie (subscript FF, FR) and M_z is the resultant moment on the wheelset from the longitudinal contact forces.

The matrices $\boldsymbol{A}$ and $\boldsymbol{B}_w$ are completed by considering the suspension arrangement by a Newton/Euler approach and arranging the resultant equations of motion accordingly. The derivative of the state vector $\dot{\boldsymbol{x}}$ contains either differential velocities (which are directly mapped to existing states) or acceleration terms that resolve from the equations of motion. $\boldsymbol{C}$ is formed to map the state definitions to measurements.

The Kalman Bucy filter is typically arranged as a closed loop observer that updates state estimates by using a linear system model to approximate the changing values, then adjusts these approximations based on any errors observed between the measured and estimated outputs. As there are no controlled inputs to the system, the equation set becomes:

$$\dot{\hat{\boldsymbol{x}}} = \boldsymbol{A}\hat{\boldsymbol{x}} + \boldsymbol{K}\left(z - \boldsymbol{C}\hat{\boldsymbol{x}}\right), \tag{5.40}$$

where $\hat{x}$ represents the state estimates.

To estimate the contact forces, the states of the system are augmented to include the unknown input variables as system states, essentially defining them as unmeasured state variables. The augmented state vector becomes:

$$x_a = \begin{bmatrix} x \\ F_{yFF} \\ M_{zFF} \\ F_{yFR} \\ M_{zFR} \end{bmatrix}. \tag{5.41}$$

The process of tuning is iterative but due to the causality of state estimation this can be done in stages. Adjustments to the weighting matrices are made to assign greater certainty to the measured states until the filter accurately represents these variables, by comparison against observed and measured outputs. It is important that these states show good agreement as they form the basis for the estimation of the creep force. Following this, adjustments can be made to the weighting relating to the contact forces (by ensuring they have less 'certainty' from measurement and rely on the obtained mathematical model) until the filter best approximates these. The resultant order of the estimator is quite large, so it is useful to group weightings to have the same values for simplicity in tuning.

The filter assumes that all the state variables are output as a measured set z. To estimate the contact forces, the state variable set is augmented to include the forces and moments acting on the wheelsets, and the filter is tuned to estimate these augmented variables, essentially forming a disturbance estimator. For example, in a quarter vehicle model, using only a primary suspension as in [16], the measured variables are the wheelset lateral and yaw positions and velocities:

$$z = \left[y_{FF}, \dot{y}_{FF}, \psi_{FF}, \dot{\psi}_{FF} \right]^T. \tag{5.42}$$

The subscript FF was used to denote the leading (front) wheelset of the leading (front) bogie.

The augmented state matrix includes the lateral contact forces F_y and the resultant moment acting on the wheelset about the 'z' axis due to the different in longitudinal forces on each contact point of the wheelset M_z:

$$x_a = \left[y_{FF}, \dot{y}_{FF}, \psi_{FF}, \dot{\psi}_{FF}, F_{y_{FF}}, M_{z_{FF}} \right]^T. \tag{5.43}$$

The A and B matrices therefore contain a dynamic model of the suspension system formed through first principal analysis or machine learning.

Model-based methods such as this typically focus on contact force estimation as the filter is unable to differentiate between forces due to creep, and those due to the geometric interaction of the wheel and rail [17]. However, the estimated forces provide extra information from which the coefficient of friction can be inferred by post processing the force estimation and using comparison with captured dynamic data such as wheelset velocity, slip and/or accelerations [18, 19].

5.3.3 Case Study: Model-Based Estimation for Low Adhesion Detection

Our acknowledgements go to the Technical Strategy Leadership Group (TSLG – RSSB UK), Transport for London for supporting this work and to the University of Sheffield as project partners. They also go to DeltaRail for providing simulation test data and assisting in the generation of the linear model for use in the model-based estimator.

This initial case study targeted the use of a vehicle model based on a British class 158 DMU (Deisel Multiple Unit), a small commuter vehicle with contemporary suspension design. The test subject was a high-fidelity, non-linear, multi-body simulation of the vehicle running in VAMPIRE to provide simulated measurement outputs and controllable adhesion conditions. VAMPIRE is a well-validated specialist rail-vehicle dynamics modelling software and provided the closest means of obtaining vehicle dynamic data short of full-scale track testing of a fully instrumented vehicle with modified adhesion conditions. Using a Model Based Physics (MBP) simulation is advantageous here as it enables a numerical comparison of the contact force estimation which is very difficult to obtain in practice. All data is simulated using non-linear contact modelling assuming interactions between a 113A Railhead and a P8 wheel. This study used four adhesion levels that represent differing risk levels of operation to operators (Table 5.6).

The VAMPIRE model is subject to a transit along a typical section of UK track under varying adhesion conditions. The irregularities modelled are statistically similar to those of a high-speed rail link found in the UK; sufficient to excite the suspension dynamics across its key dynamic range. Each simulated test run was

TABLE 5.6
Definition of coefficient of friction for operating conditions

Coefficient of Friction - ¼	Descriptor	comment
0.56	'Dry'	No impact on operation
0.32	'Wet'	No impact on operation
0.072	'Low'	Mitigation by defensive driving
0.038	'Very low'	Little mitigation action possible to reduce risk

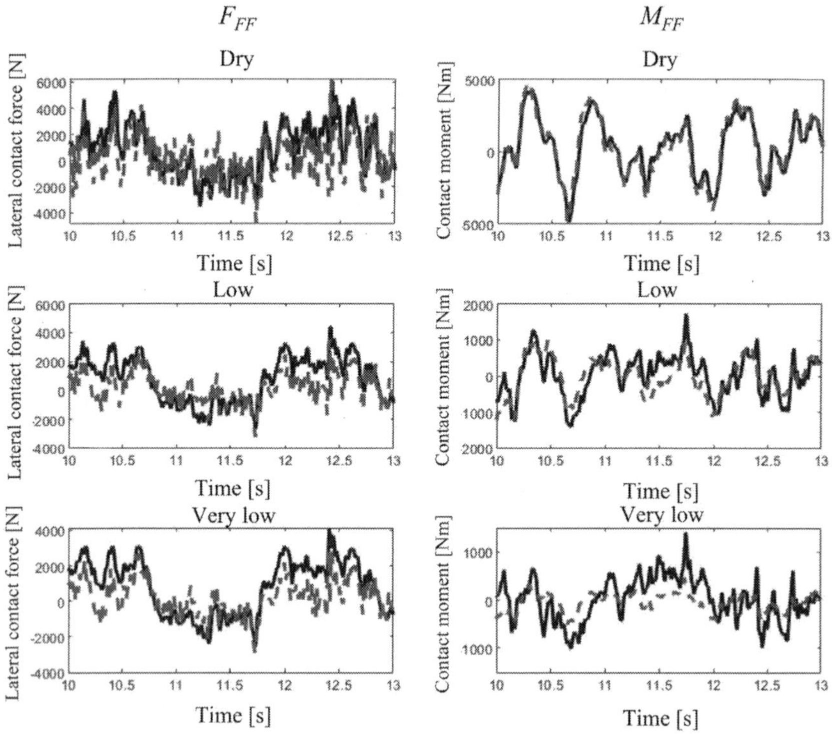

FIGURE 5.38 Force/Moment estimation in normal running case study.

conducted with a vehicle speed of 200 km/h and lasts for 60 seconds. Again, the data is collected as if it were measured data and subject to processing through the model-based estimator. Figure 5.38 shows the result of the contact force estimation for three adhesion conditions. The figure focuses on a three second time interval to better observe estimation results.

Figure 5.38 shows that the contact moment on the wheelset is best estimated by this method, and that the general magnitude of moment decreases in line with the coefficient of friction. The best results to obtain an estimate of adhesion $\hat{\mu}$ was to normalise a 5 second moving RMS of moment estimation with an equivalent 5 second moving RMS of wheelset rotational acceleration with the intention of scaling the moment estimation with an indicator of track irregularity magnitude. These numerical values are regressed against known experimental data to create a lookup table for $\hat{\mu}$. Figure 5.39 shows two experiments where the known coefficient of friction reduces from normal ('Dry') operation to low levels and the estimator establishes an estimate of the change in adhesion.

A source of inaccuracy in this estimation was identified as the uncertainty of the linear model used to form the A matrix in the Kalman Filter. The model makes linear assumptions about the movement of the running gear from which the contact

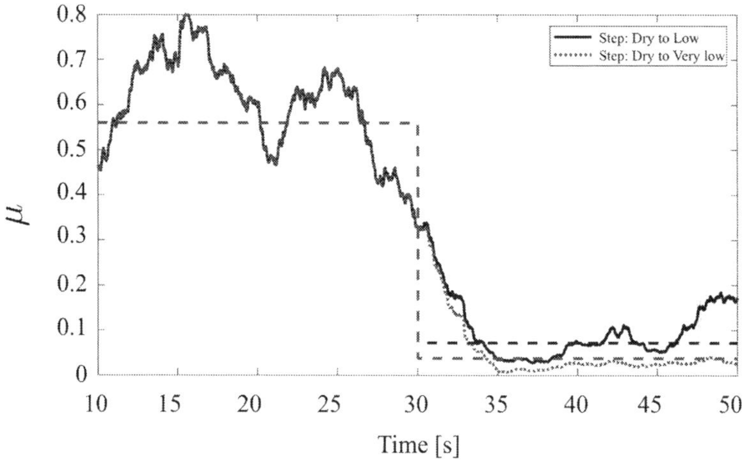

FIGURE 5.39 Step change in adhesion level during running.

forces are estimated. Model uncertainty due to the complex geometries and non-linear properties of suspension components means that estimations will be at their greatest error typically at low adhesion levels. This is when the stimulus of the system is at its smallest due to the reduction in forces acting on the wheelset. To improve estimation at the critical low adhesion conditions, better models are required at these operating conditions.

A potential solution to this problem is to use systems identification to best approximate a linear model of the running gear at low adhesion conditions, in place of a first principal analysis. The following case study details results obtained when using a SIMPACK multi-body simulation model of a D78 Rolling stock unit used in the London commuter network. SIMPACK is a well verified tool used widely in the rail industry for vehicle studies and is effective at simulating the complex wheel/rail interfaces that are necessary in this study. Sensor measurements of inertial data are mimicked using data generated from the simulation. Full details of this study are reported in [20].

The system description for the Kalman filter is populated by completing a systems identification exercise using the least squares method (see Appendix A2.3) for each state variable derivative term defined in $\dot{x}_a$. The variables for the systems identification exercise are output from experimental data generated from the simulation model. Figure 5.40 shows an example of verification data for two variables. Typically, the longitudinal variables outperformed the lateral variables throughout the system.

The rows of the state description describing the augmented states of Force and Moment can also be populated using this method. As an MBP simulation is used, the time histories of contact forces can be extracted from the simulation as variables, and subject to a systems identification process against the state variables,

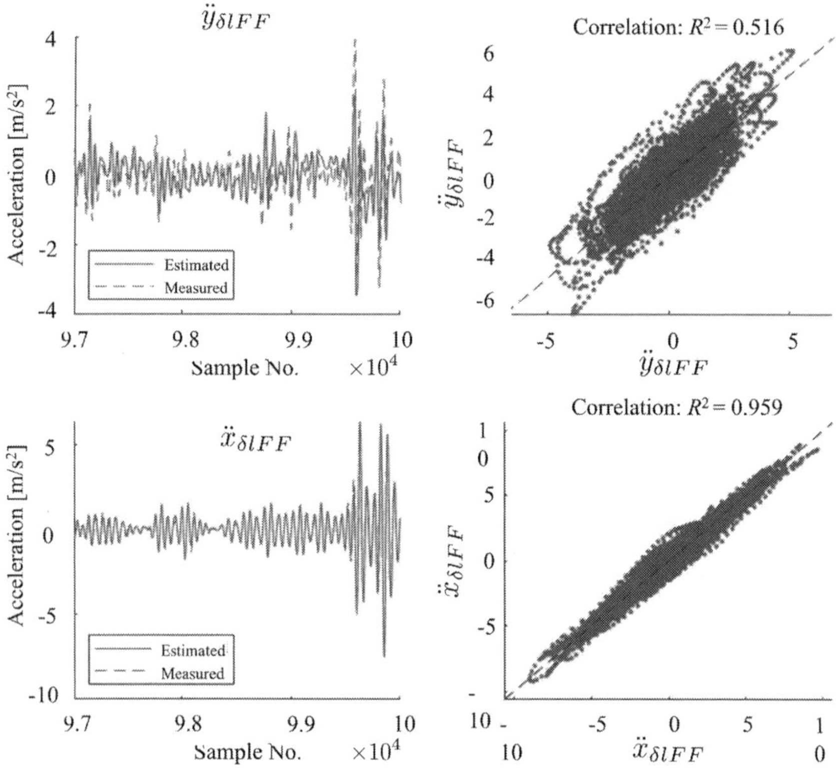

FIGURE 5.40 Example of systems identification results.

as with the inertial states. This provides the estimator with an approximate model of how the contact forces and running gear dynamics are coupled. The inclusion of this part of the estimator will lead to better attribution of the estimated augmented states to the system states in the estimation process. Furthermore, it is also possible to include the creep as augmented states that in turn can be estimated. Estimating both creep and creep force allows for a real time formulation of the creep curves experienced during running, and the subsequent option to use these estimated properties as lookup references to estimate adhesion from known or theoretical curves.

Figure 5.41 shows the results of estimating adhesion using only inertial data. The estimator outputs creep and creep force, which are each then subject to a 5 second moving RMS. The values of creep and creep force are then interpolated in a look up table formed from established creep curves to provide an estimated coefficient of friction.

The results show that the estimator performs a reasonable task of identifying the operating condition of the vehicle, even if the numerical values lack absolute accuracy. To enable better identification of low adhesion conditions, the systems

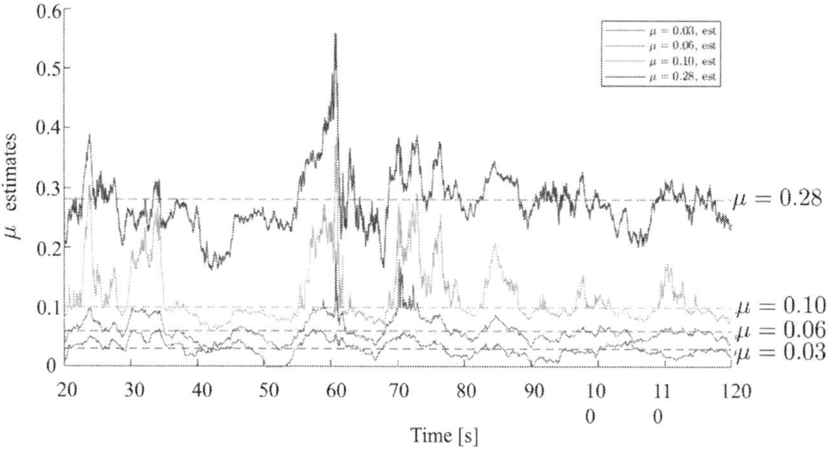

FIGURE 5.41 Coefficient of friction estimation using look-up table for μ.

identification to establish a suspension model was completed using data of similar adhesion conditions so that the accuracy of estimation output is deliberately biased towards effectiveness at operating conditions in low adhesion.

5.3.4 Case Study: Data-Driven (Model-Free) Approach to LAD (Network Rail)

This FNN approach presented in [21] is the result of collaborative research between Loughborough University, The University of Sheffield, and Hitachi Rail UK, sponsored by Network Rail UK [22]. The intention of the project was to artificially change the coefficient of friction of a section of test track, and collect inertial data from a rail vehicle operating over these changing conditions. Post processing of this data would determine if a difference in the vehicle dynamics could be observed due to the changing adhesion, and also if the inertial readings could be used to estimate the coefficient of friction. This could then be verified if the operating condition is recorded for each test conducted.

The vehicle used was a Multi-Purpose Vehicle (MPV) owned by Network Rail that comprises of 2 cars each with a pair of four-wheel bogies. Only one of the cars was instrumented, and it is shown in Figure 5.42. Instrumentation of the vehicle was carried out by Perpetuum, now part of Hitachi Rail. The vehicle was instrumented by accelerometers and potentiometers on all four wheelsets and two bogies. The left and right axleboxes of the wheelsets were instrumented with 3-axis Dytran 7533A4 accelerometers. Figure 5.42 also shows the fixing of the accelerometers to the axlebox on a modified keeper plate in a 3D printed housing.

Experimental conditions were created [23] and verified by the use of adhesion modifiers on an 800m section of a test track at Tuxford, UK, a Network Rail owned facility. The obtained coefficient of friction in each test case was measured

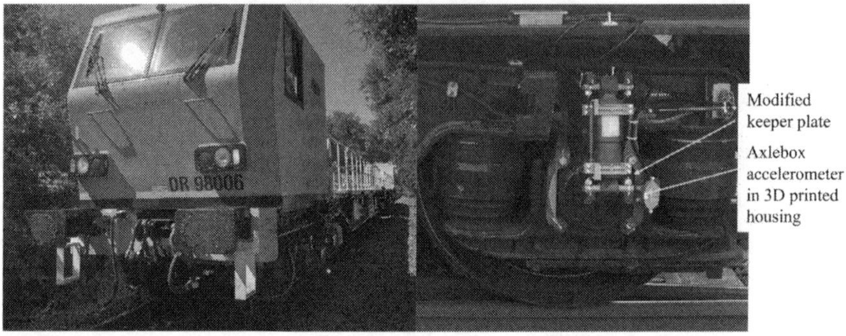

FIGURE 5.42 MPV and sensor installation on the axlebox and across the primary suspension.

TABLE 5.7
Summary of available test cases

Speed	16 mph	26 mph	40 mph	60 mph
Straight track				
Dry	0.31	0.29	0.31, 0.37	0.3, 0.34, 0.35
FM	0.14, 0.17	0.19	0.17	0.16
Wet	0.161	0.19	0.24, 0.24	0.27
Detergent	0.18, 0.17	0.21	0.17, 0.22	0.24
Paper tape	0.16, 0.2	0.07, 0.16, 0.22	0.1	0.11, 0.19
Curved track				
Dry	0.28	0.32	0.31	0.36
FM	0.13, 0.26	0.26	0.27	0.27
Wet	0.35, 0.23	0.19	0.25	-
Detergent	0.28	0.25, 0.28	0.28, 0.32	0.28

at select points along the test section to verify the experimental condition, and all test conditions are summarised in Table 5.7.

A model-based approach to adhesion estimation in this case proved troublesome as it was difficult to obtain a sufficiently representative linear model of the rubber-block suspension of the MPV. Instead, as a large amount of data was collected with known test conditions, it seemed reasonable that a model-less approach could be taken where a neural network could be trained on the inertial data to recognise changes in the coefficient of friction [21]. Options of neural network structures are available with varying attributes, but the most successful implementation makes use of a Deep Convoluted Neural Network (DCNN). Instead of operating on time-domain acceleration signals, a Short-Time Fourier Transform (STFT) is applied to convert them from 1D to 2D signals for convolution operation. The frequency-domain approach is chosen since it is more powerful in identifying frequency features and spectral energy associated with specific friction/adhesion levels. In this case acceleration measurements from the axleboxes were the primary data inputs to the network, following some simple filtering before a STFT to obtain the 2D data.

To protect from over-fitting the experimental data is partitioned into three sections for training, validation and test. The training and validation data come from the same test run, but are partitioned by time such that validation (and subsequent tuning) occurs on data not seen during training. Complete data sets are reserved for the final stages for test purposes.

The architecture of the DCNN consists of 4 Convolutional Layers (CLs), denoted as CL1, CL2, CL3, and CL4 respectively, and a Fully Connected Layer (FCL). This architecture belongs to an end-to-end discriminative model, where feature extraction and classification are performed simultaneously. Discriminative models have also been shown to be more accurate than generative models [24]. The outputs of the CLs are referred to as Feature Vectors (FVs). Commonly, each of the CLs is followed by a ReLU and a pooling layer, where the former helps reduce overfitting, and the latter reduces the dimensions of the FVs and size of the network for faster training and implementation. Average Pooling Layers (APLs) are used for pooling, which lead to more consistent estimations. There is also a dropout layer with a dropout factor of 0.5 between each ReLU and APL layer to reduce overfitting. The convolution filter head in each CL is 150. The length of convolution filters is 3; a stride of 1 and the same padding are employed. The size and stride of the APLs is 2, meaning that the FVs' sizes after each APL are halved. The FVs from the final CL are then flattened by the FCL, which also connects every element of the final FV to a single neuron, as in an MLP. The STFT inputs are also normalized to [-1, 1] first using their respective maximum and minimum values for improved training. The architecture of the MC-DCNN is shown below in Figure 5.43, where the ReLU layers and APLs are not shown, and μ_{est} is the estimated friction coefficient after training.

The number of convolutional filter heads used decides the number of features to extract from the previous layer. The consideration is to use a sufficient number of convolution filters such that all existing features are extracted from the inputs. However, in addition to increasing network complexity, using too many filter heads can lead to overfitting, long training time and poor generalization, especially if the data does not include all possible running conditions. A higher number is chosen to ensure that this is not a limiting factor. The number of CLs depends on the hierarchical relationship between the input and the output, and the number of classes. After some training and validation studies, it was found that increasing the number of CLs improves the consistency of the estimation between different training instances, meaning increased robustness against random network parameter initializations. For all speed cases, four CLs was found to be sufficient after some tuning, and adding more layers hugely increased the training time without improving the training error.

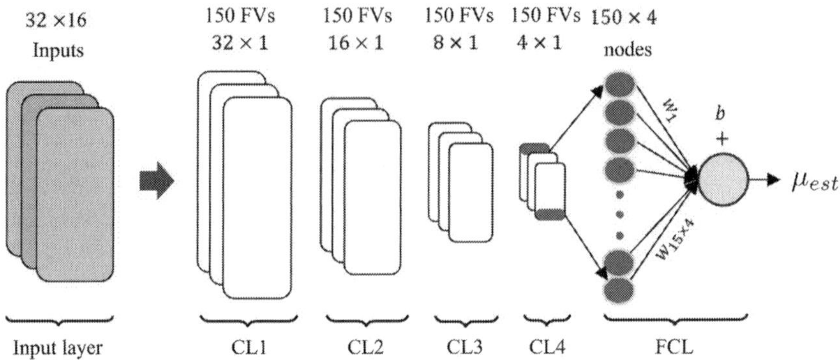

FIGURE 5.43 Architecture of the MC-DCNN.

TABLE 5.8
The average $\bar{\mu}_{est}$, average RMSEs, average MAE, and the SD of MAEs from 10 MC-DCNNs

Measurement set	16 mph	26 mph	40 mph	60 mph
Straight track				
μ	0.17	0.16	0.22	0.3
Average $\bar{\mu}_{est}$	0.247	0.179	0.235	0.304
Average RMSE	0.0774	0.0209	0.0165	0.0083
Average MAE	0.0774	0.0204	0.0164	0.0068
SD (MAE)	0.0125	0.0221	0.0085	0.0037
Curved track				
μ	0.23	0.25	0.28	0.27
Average $\bar{\mu}_{est}$	0.246	0.259	0.25	0.338
Average RMSE	0.0224	0.0095	0.0287	0.0685
Average MAE	0.0219	0.0094	0.0287	0.0684
SD (MAE)	0.0192	0.0132	0.0041	0.0043

Table 5.8 shows the results of the DCNN against unseen test data across different speed conditions, and through both straight and curved track sections.

The estimator performs very well across both changing speed and changing adhesion. Although this method is promising, there is some consideration to be made about the application to a 'general' case of estimation in that learned systems are only really verified within the bounds of the training data. For in-service implementation, training data should be established (through on-track testing or augmented by simulation data) for much wider operating conditions of adhesion conditions, and include different track profiles and wear conditions of wheels.

REFERENCES

1. A. Matsumoto, Y. Sato, H. Ohno, M. Shimizu, J. Kurihara, T. Saitou, Y. Michitsuji, R. Matsui, M. Tanimoto, M. Mizuno, Actual states of wheel/rail contact forces and friction on sharp curves - Continuous monitoring from in-service trains and numerical simulations, *Wear*, 314, 189–197, 2014.
2. A. Matsumoto, Y. Michitsuji, Y. Ichiyanagi, Y. Sato, H. Ohno, M. Tanimoto, A. Iwamoto, T. Nakai, Safety measures against flange-climb derailment in sharp curve - considering friction coefficient between wheel and rail, *Wear*, 432–433, 202931, 1-16, 2019.
3. Y. Ichiyanagi, Y. Michitsuji, A. Matsumoto, Y. Sato, H. Ohno, S. Ogata, M. Tanimoto, A. Iwamoto, T. Fukushima, T. Nakai, Running position matching for the monitoring bogie and temporal subtraction analysis of derailment coefficient, *Mechanical Engineering Journal*, 6(2), 1–12, 2019.
4. M. Odashima, S. Azami, Y. Naganuma, H. Mori, H. Tsunashima, Track geometry estimation of a conventional railway from car-body acceleration measurement, *Mechanical Engineering Journal*, 4(1), 1–12, 2017.
5. M. Ray, M. Anghileri, M. Mongiardini, Comparison of validation metrics using repeated full-scale automobile crash tests, In: Schrefler, B.A., Perego, U. (eds) *Proceedings of the 8th World Congress on Computational Mechanics,* The International Association for Computational Mechanics, 2008.
6. H. Tsunashima, N. Yagura, Railway track irregularity estimation using car body vibration: A data-driven approach for regional railway, *Vibration*, 7, 928–948, 2024.
7. J. Wang, *An Intuitive Tutorial to Gaussian Processes Regression*, arXiv 2009, arXiv:2009.10862.
8. H. Claus, W. Schiehlen, Modeling and simulation of railway bogie structural vibrations, *Vehicles Systems Dynamics*, 29(Suppl. 1), 538–552, 1998.
9. D.H. Wang, W.H. Liao, Semi-active suspension systems for railway vehicles using magnetorheological dampers, *Part I: System Integration and Modelling*, *Vehicles Systems Dynamics*, 47, 1305–1325, 2009.
10. A. Berawi, Improving railway track maintenance using Power Spectral Density (PSD). Doctoral Dissertation, University of The University of Porto, August 2013. https://api.semanticscholar.org/CorpusID:111822997 (accessed on 1 July 2024).
11. M. Spiryagin, C. Cole, Y.Q. Sun, Adhesion estimation and its implementation for traction control of locomotives, *International Journal of Rail Transportation*, 2(3), 187–204, 2014.
12. S. Shrestha, M. Spiryagin, Q. Wu, Friction condition characterization for rail vehicle advanced braking systems, *Mechanical Systems and Signal Processing*, 134, 106324, 2019.
13. L. Jianhua, L. Linfan, H. Jing, Z. Changfan, Z. Kaihui, Wheel/rail adhesion state identification of heavy-haul locomotives based on particle swarm optimization and Kernel extreme learning machine, *Journal of Advanced Transportation*, Volume 2020, Article ID 8136939, 1–6, 2020.
14. M. Malvezzi, L. Pugi, S. Papini, A. Rindi, P. Toni, Identification of a wheel–rail adhesion coefficient from experimental data during braking tests, *Proceedings of the Institution of Mechanical Engineers, Part F: Journal of Rail and Rapid Transit*, 227(2), 128–139, 2013.

15. S. Shrestha, A. Koirala, M. Spiryagin, Q. Wu, Wheel-rail interface condition estimation via acoustic sensors, *Proceedings of the 2020 Joint Rail Conference*, JRC2020-8037, V001T03A005, American Society of Mechanical Engineers, 2020.
16. P.D. Hubbard, C.P. Ward, R. Dixon R.M. Goodall, Real time detection of low adhesion in the wheel/rail contact, *Proceedings of the Institution of Mechanical Engineers, Part F: Journal of Rail and Rapid Transit*, 227(6), 623–634, 2013.
17. P.D. Hubbard, C.P. Ward, R. Dixon, R.M. Goodall, Models for estimation of creep forces in the wheel/rail contact under varying adhesion levels, *Vehicle System Dynamics: International Journal of Vehicle Mechanics and Mobility*, Sup 1(IAVSD Proceedings Supplement), 370–386, 2014.
18. S. Shrestha, Q. Wu, S. Maksym, Review of adhesion estimation approaches for rail vehicles, *International Journal of Rail Transportation*, 7(2), 79–102, 2019.
19. S. Meymand, A. Keylin, M. Ahmadian, A survey of wheel–rail contact models for rail vehicles, *Vehicle System Dynamics*, 54(3), 386–428, 2016.
20. P. Hubbard, T. Harrison, C. Ward, B. Abduraxman, Creep slope estimation for assessing adhesion in the wheel/rail contact, *IET Intelligent Transport Systems*, 18, 1931–1942, 2024.
21. B. Abduraxman, P. Hubbard, T. Harrison, C. Ward, D. Fletcher, R. Lewis, B. White, Acceleration-based friction coefficient estimation of a rail vehicle using feedforward NN: Validation with track measurements, *Vehicle System Dynamics*, 62(12), 3235–3254, 2024.
22. T. Harrison, B. Abduraxman, P. Hubbard, B. White, D. Fletcher, R. Lewis, K. Chandrashekar, D. Vincent, S. Chaney, M. Burstow, E. Cockroft, Creation of and measurement of low adhesion conditions for the development of on-train low adhesion detection equipment, In: Huang, W., Ahmadian, M. (eds) *Advances in Dynamics of Vehicles on Roads and Tracks III. IAVSD 2023. Lecture Notes in Mechanical Engineering*, Springer, 2025.
23. B. White, R. Lewis, D. Fletcher, T. Harrison, P. Hubbard, C. Ward, Rail-wheel friction quantification and its variability under lab and field trial conditions, *Proceedings of the Institution of Mechanical Engineers, Part F: Journal of Rail and Rapid Transit*, 238(5), 569–579, 2024.
24. H.I. Fawaz, G. Forestier, J. Weber, L. Idoumghar, P.-A. Muller, Deep learning for time series classification: A review, *Data Mining and Knowledge Discovery*, 33, 917–963, 2018.

Appendices

A1 STATE-SPACE REPRESENTATION OF DYNAMIC SYSTEMS

A1.1 Representation of Dynamic Systems in Continuous Time

In this section, a state-space representation is derived as the representation form of the dynamic system for formulating the Kalman filter (KF).

Consider the mechanical system of mass, spring and damper systems shown in Figure A1.1. The equation of motion is expressed as

$$\ddot{y}(t) = -\frac{c}{m}\dot{y}(t) - \frac{k}{m}y(t) + \frac{1}{m}u(t), \tag{A1.1}$$

where m is the mass, c is the damping coefficient, k is the spring constant, $u(t)$ is the force acting on the object (input) and $y(t)$ is the position of the object (output).

Now, as state variables, we choose the position and velocity of the object as follows:

$$x_1(t) = y(t),$$

$$x_2(t) = \dot{y}(t) = \dot{x}_1(t).$$

Then, the following first-order simultaneous differential equations are obtained:

$$\dot{x}_1(t) = x_2(t), \tag{A1.2}$$

$$\dot{x}_2(t) = -\frac{c}{m}x_2(t) - \frac{k}{m}x_1(t) + \frac{1}{m}u(t). \tag{A1.3}$$

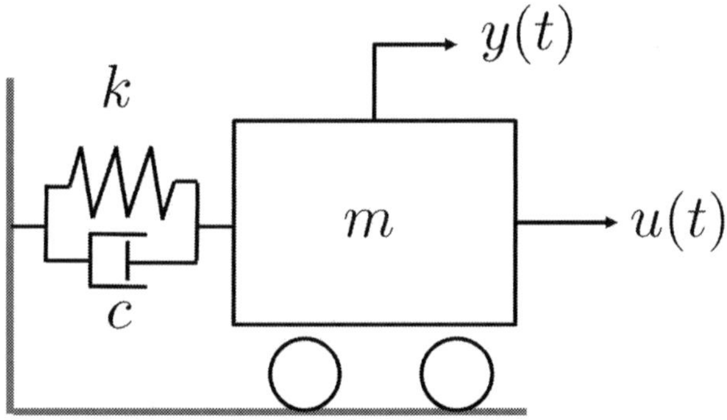

FIGURE A1.1 Mass, spring and damper mechanical systems.

Equations (A1.2) and (A1.3) can be expressed in matrix form as follows:

$$\underbrace{\begin{bmatrix} \dot{x}_1(t) \\ \dot{x}_2(t) \end{bmatrix}}_{\dot{x}(t)} = \underbrace{\begin{bmatrix} 0 & 1 \\ -\dfrac{k}{m} & -\dfrac{c}{m} \end{bmatrix}}_{A} \underbrace{\begin{bmatrix} x_1(t) \\ x_2(t) \end{bmatrix}}_{x(t)} + \underbrace{\begin{bmatrix} 0 \\ \dfrac{1}{m} \end{bmatrix}}_{B} u(t). \tag{A1.4}$$

where $\boldsymbol{x}(t)$ is called the state vector.

To measure the position of an object, the following state quantities need to be determined:

$$z(t) = x_1(t) \tag{A1.5}$$

Equation (A1.5) can be expressed in the following matrix form:

$$z(t) = \underbrace{\begin{bmatrix} 1 & 0 \end{bmatrix}}_{C} \underbrace{\begin{bmatrix} x_1(t) \\ x_2(t) \end{bmatrix}}_{x(t)}. \tag{A1.6}$$

In general, a dynamic system can be described as follows:

$$\dot{\boldsymbol{x}}(t) = \boldsymbol{A}\boldsymbol{x}(t) + \boldsymbol{B}\boldsymbol{u}(t), \tag{A1.7}$$

$$z(t) = \boldsymbol{C}\boldsymbol{x}(t), \tag{A1.8}$$

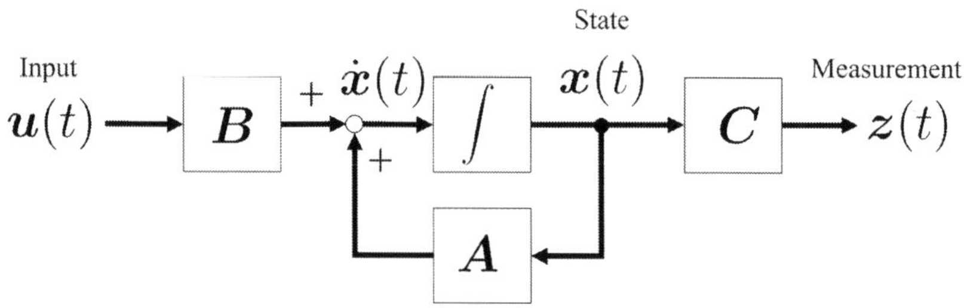

FIGURE A1.2 State space representation.

where

$$A=\left[a_{ij}\right], B=\left[b_{ij}\right], C=\left[c_{ij}\right],$$

$$x(t)=\begin{bmatrix} x_1(t) \\ x_2(t) \\ \vdots \\ x_n(t) \end{bmatrix}, u(t)=\begin{bmatrix} u_1(t) \\ u_2(t) \\ \vdots \\ u_m(t) \end{bmatrix}, z(t)=\begin{bmatrix} z_1(t) \\ z_2(t) \\ \vdots \\ z_l(t) \end{bmatrix}, \dot{x}(t)=\begin{bmatrix} \dot{x}_1(t) \\ \dot{x}_2(t) \\ \vdots \\ \dot{x}_n(t) \end{bmatrix}.$$

Equations (A1.7) and (A1.8) are called the state-space representation of the system and are called the state equation and the measurement equation, respectively. They can also be represented using a block diagram as shown in Figure A1.2.

A1.2 Representation of Dynamic Systems in Discrete Time

Equations (A1.7) and (A1.8) are expressions for dynamic systems in continuous time. This section describes the representation of the dynamic system in discrete time.

The equations of state and measurement of a dynamic system in continuous time are described below.

$$\dot{x}(t)=Ax(t)+Bu(t)+\tilde{w}(t), \tag{A1.9}$$

$$z(t)=Cx(t)+\tilde{v}(t). \tag{A1.10}$$

The equation of state in Equation (A1.9) is the addition of system noise (also called plant or process noise) $\tilde{w}(t)$ to Equation (A1.7). The system noise represents the inaccuracy of Equation (A1.7). The measurement Equation (A1.10) is the addition of the sensor noise $\tilde{v}(t)$ to Equation (A1.8).

Now, it is assumed that $\tilde{\boldsymbol{w}}(t)$ is white Gaussian noise with zero mean and a covariance matrix $\boldsymbol{W}(t)$. It's also assumed that $\tilde{\boldsymbol{v}}(t)$ is white Gaussian noise with zero mean and covariance matrix $\boldsymbol{V}(t)$.

Solving the state equation in Equation (A1.9) yields

$$\boldsymbol{x}(t) = e^{A(t-t_0)}\boldsymbol{x}(t_0) + \int_{t_0}^{t} e^{A(t-\alpha)}\{\boldsymbol{Bu}(\alpha) + \tilde{\boldsymbol{w}}(\alpha)\}d\alpha. \tag{A1.11}$$

where $\boldsymbol{x}(t_0)$ is the initial value and $e^{A(t-t_0)}$ is called the state transition matrix from time t_0 to time t. e^{At} is the matrix exponential function, defined by

$$e^{At} = \sum_{k=0}^{\infty}\frac{(At)^k}{k!} = I + \frac{At}{1!} + \frac{(At)^2}{2!} + \frac{(At)^3}{3!} + \cdots, \tag{A1.12}$$

where $\boldsymbol{I}$ is an $n \times n$ identity matrix.

Now, the input $\boldsymbol{u}(t)$ is assumed to be constant at $t_{k-1} \leq t < t_k$.

$$\boldsymbol{u}(t) = \boldsymbol{u}(t_{k-1}) \qquad t_{k-1} \leq t < t_k. \tag{A1.13}$$

The state $\boldsymbol{x}(t_k)$ at sampling time t_k can be obtained by setting $t = t_k$ and $t_0 = t_{k-1}$ in Equation (A1.11) as follows:

$$\boldsymbol{x}(t_k) = \boldsymbol{Fx}(t_{k-1}) + \boldsymbol{Gu}(t_{k-1}) + \boldsymbol{w}(t_{k-1}). \tag{A1.14}$$

For simplicity, the following equation is used here:

$$\boldsymbol{x}(k) = \boldsymbol{Fx}(k-1) + \boldsymbol{Gu}(k-1) + \boldsymbol{w}(k-1). \tag{A1.15}$$

If the sampling period is written as τ ($=t_k - t_{k-1}$), the matrices $\boldsymbol{F}$ and $\boldsymbol{G}$ are described by

$$F = e^{A(t_k - t_{k-1})} = e^{A\tau} = \boldsymbol{I} + A\tau + \frac{A^2\tau^2}{2!} + \cdots, \tag{A1.16}$$

$$\boldsymbol{G} = \int_0^{\tau} e^{A(\tau-t)}\boldsymbol{B}dt = \boldsymbol{B}\tau + \frac{\boldsymbol{AB}\tau^2}{2!} + \cdots. \tag{A1.17}$$

The covariance matrix for the system noise $\boldsymbol{w}(k-1)$ in discrete time is

$$\boldsymbol{Q}=\int_0^\tau e^{A(\tau-t)}\mathbf{W}(t)\left\{e^{A(\tau-t)}\right\}^T dt=\mathbf{W}(t)\,\tau+\frac{\left(\boldsymbol{AW}(t)+\boldsymbol{W}(t)\boldsymbol{A}^T\right)\tau^2}{2!}+\cdots. \quad \text{(A1.18)}$$

If the sampling time τ is sufficiently small and higher-order terms above τ^2 can be neglected compared to terms in τ, then

$$\boldsymbol{F}=\boldsymbol{I}+\boldsymbol{A}\tau, \quad \text{(A1.19)}$$

$$\boldsymbol{G}=\boldsymbol{B}\tau, \quad \text{(A1.20)}$$

$$\boldsymbol{Q}=\mathbf{W}(t)\,\tau. \quad \text{(A1.21)}$$

The measurement equation in discrete time can be expressed as

$$z(k)=\boldsymbol{Hx}(k)+\boldsymbol{v}(k). \quad \text{(A1.22)}$$

We now compare this with the measurement equation in continuous time in Equation (A1.8), where $\boldsymbol{H}=\boldsymbol{C}$ and the covariance matrix of $\boldsymbol{v}(k)$ is

$$\boldsymbol{R}=\frac{\boldsymbol{V}(t)}{\tau}. \quad \text{(A1.23)}$$

Example 1.1
Consider a mass in one-dimensional motion at a constant velocity. If the position of the mass is $x(t)$, the motion of the mass can be expressed as follows:

$$\ddot{x}(t)=\tilde{w}(t), \quad \text{(A1.24)}$$

where $\tilde{w}(t)$ represents the acceleration disturbance and is white Gaussian noise with a mean of zero and a variance of q.

The equation of state for the system in continuous time is

$$\begin{bmatrix}\dot{x}(t)\\ \ddot{x}(t)\end{bmatrix}=\begin{bmatrix}0 & 1\\ 0 & 0\end{bmatrix}\begin{bmatrix}x(t)\\ \dot{x}(t)\end{bmatrix}+\begin{bmatrix}0\\ \tilde{w}(t)\end{bmatrix}. \quad \text{(A1.25)}$$

Equation (A1.21) can be expressed as

$$\dot{\boldsymbol{x}}(t) = \boldsymbol{A}\boldsymbol{x}(t) + \tilde{\boldsymbol{w}}(t), \tag{A1.26}$$

where

$$\boldsymbol{x}(t) = \begin{bmatrix} x(t) \\ \dot{x}(t) \end{bmatrix} \quad \boldsymbol{A} = \begin{bmatrix} 0 & 1 \\ 0 & 0 \end{bmatrix} \quad \tilde{\boldsymbol{w}}(t) = \begin{bmatrix} 0 \\ \tilde{w}(t) \end{bmatrix}.$$

Based on Equation (A1.22), the state equation in discrete time is

$$\boldsymbol{x}(k) = \boldsymbol{F}\boldsymbol{x}(k-1) + \boldsymbol{w}(k-1). \tag{A1.27}$$

Using Equation (A1.16), the matrix $\boldsymbol{F}$ becomes

$$\boldsymbol{F} = e^{A\tau} = \begin{bmatrix} 1 & \tau \\ 0 & 1 \end{bmatrix} \tag{A1.28}$$

From Equation (A1.18), the covariance matrix for the system noise $\boldsymbol{w}(k-1)$ becomes

$$\boldsymbol{Q} = \begin{bmatrix} \frac{1}{3}\tau^3 & \frac{1}{2}\tau^2 \\ \frac{1}{2}\tau^2 & \tau \end{bmatrix} q. \tag{A1.29}$$

The representation of dynamic systems in discrete time is summarised below.

The state equation is

$$\boldsymbol{x}(k) = \boldsymbol{F}\boldsymbol{x}(k-1) + \boldsymbol{G}\boldsymbol{u}(k-1) + \boldsymbol{w}(k-1), \tag{A1.30}$$

where $\boldsymbol{w}(k-1)$ is white Gaussian noise with a zero mean and a covariance matrix $\boldsymbol{Q}$.

The measurement equation is

$$z(k) = \boldsymbol{H}\boldsymbol{x}(k) + \boldsymbol{v}(k), \tag{A1.31}$$

where $\boldsymbol{v}(k)$ is white Gaussian noise with a zero mean and a covariance matrix $\boldsymbol{R}$.

Note that, in Equations (A1.26) and (A1.27), $\boldsymbol{F}$, $\boldsymbol{G}$ and $\boldsymbol{H}$ are expressed as constant matrices, that is, as time-invariant systems, but they can also be written as time-invariant systems as follows.

$$\boldsymbol{x}(k)=\boldsymbol{F}(k-1)\boldsymbol{x}(k-1)+\boldsymbol{G}(k-1)\boldsymbol{u}(k-1)+\boldsymbol{w}(k-1), \tag{A1.32}$$

$$z(k)=\boldsymbol{H}(k)\boldsymbol{x}(k)+\boldsymbol{v}(k). \tag{A1.33}$$

A2 DISCRETE-TIME KALMAN FILTER

A2.1 Probability Distribution

A2.1.1 Expected Value and Variance

First, consider the case of discrete random variables. The probability that the random variable X takes x_i is defined as

$$P(X=x_i)=p(x_i),(i=1,2,\cdots,n), \tag{A2.1}$$

where

$$\sum_{i=1}^{n}p(x_i)=1\ . \tag{A2.2}$$

If the random variable takes continuous values, the following definition is used:

$$P(a\le x\le b)=\int_a^b p(x)dx\ , \tag{A2.3}$$

where $p(x)$ is called the probability density function.

The mean value is defined by the following equation:

$$\mu=E[x]=\sum_{i=1}^{n}x_i p(x_i), \tag{A2.4}$$

where $E[x]$ is called the expected value of x and is the value taken by the random variable weighted by the probability of occurrence.

If the random variable is continuous, the mean value is defined by the following equation:

$$\mu=E[x]=\int_{-\infty}^{\infty}xp(x)dx\ . \tag{A2.5}$$

The variance is also defined as follows for discrete and continuous random variables, respectively:

$$\sigma^2 = E\left[(x-\mu)^2\right] = \sum_{i=1}^{n}\left(x_i-\mu\right)^2 p\left(x_i\right), \tag{A2.6}$$

$$\sigma^2 = E\left[(x-\mu)^2\right] = \int_{-\infty}^{\infty}(x-\mu)^2 p(x)\,dx. \tag{A2.7}$$

Equation (A2.7) can be written as

$$\sigma^2 = E\left[x^2-2x\mu+\mu^2\right] = E\left[x^2\right]-2\mu^2+\mu^2 = E\left[x^2\right]-\mu^2. \tag{A2.8}$$

Now consider the case where the random variable is vector-valued. Let $\boldsymbol{x} = \left[x_1, x_2, \cdots, x_n\right]^T$ be an n-dimensional random variable vector. In this case, the mean vector μ is

$$\boldsymbol{\mu} = \begin{bmatrix} E\left[x_1\right] \\ E\left[x_2\right] \\ \vdots \\ E\left[x_n\right] \end{bmatrix} = \begin{bmatrix} \mu_1 \\ \mu_2 \\ \vdots \\ \mu_n \end{bmatrix}. \tag{A2.9}$$

The covariance matrix is also a symmetric matrix ($\sigma_{ij} = \sigma_{ji}$) expressed as

$$\boldsymbol{\Sigma} = \begin{bmatrix} \sigma_{11}^2 & \cdots & \sigma_{1n} \\ \vdots & \ddots & \vdots \\ \sigma_{n1} & \cdots & \sigma_{nn}^2 \end{bmatrix}, \tag{A2.10}$$

where

$$\sigma_{ij}^2 = E\left[\left(x_i-\mu_i\right)\left(x_j-\mu_j\right)^T\right]. \tag{A2.11}$$

It can alternatively be expressed as

$$\boldsymbol{\mu} = E\left[\boldsymbol{x}\right], \tag{A2.12}$$

$$\boldsymbol{\Sigma} = E\left[(\boldsymbol{x}-\boldsymbol{\mu})(\boldsymbol{x}-\boldsymbol{\mu})^T\right] = E\left[\boldsymbol{x}\boldsymbol{x}^T\right]-\boldsymbol{\mu}\boldsymbol{\mu}^T. \tag{A2.13}$$

A2.1.2 Normal (Gaussian) Distribution

A probability distribution where the probability density function of the random variable x is given by

$$p(x) = \frac{1}{\sqrt{2\pi\sigma^2}} \exp\left[\frac{-(x-\mu)^2}{2\sigma^2}\right]. \tag{A2.14}$$

is called a normal distribution or a Gaussian distribution, where μ is the mean value of the random variable x and σ^2 is the variance.

The normal distribution is denoted by $N(\mu, \sigma^2)$. Figure A2.1 shows the normal distribution for $\mu = 0$ and $\sigma^2 = 1, 2, 4$.

For an n-dimensional vector of random variables, the probability density function is expressed by

$$p(\boldsymbol{x}) = \frac{1}{(2\pi)^{\frac{n}{2}} |\boldsymbol{\Sigma}|^{\frac{1}{2}}} \exp\left[-\frac{1}{2}(\boldsymbol{x}-\boldsymbol{\mu})^T \boldsymbol{\Sigma}^{-1}(\boldsymbol{x}-\boldsymbol{\mu})\right]. \tag{A2.15}$$

Figure A2.2 shows an example where the random variables are two-dimensional, $x_1, x_2, \mu_1 = \mu_2 = 0, \sigma_{11}^2 = \sigma_{22}^2 = 1, \sigma_{12} = \sigma_{21} = 0$.

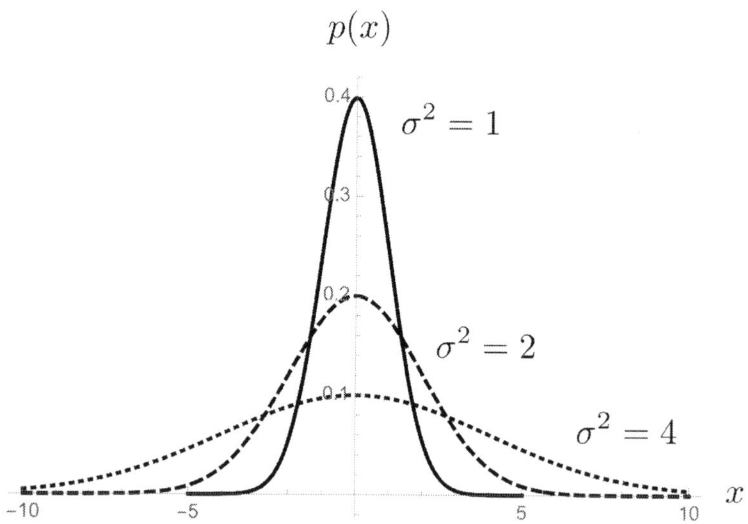

FIGURE A2.1 Probability density function of the normal distribution.

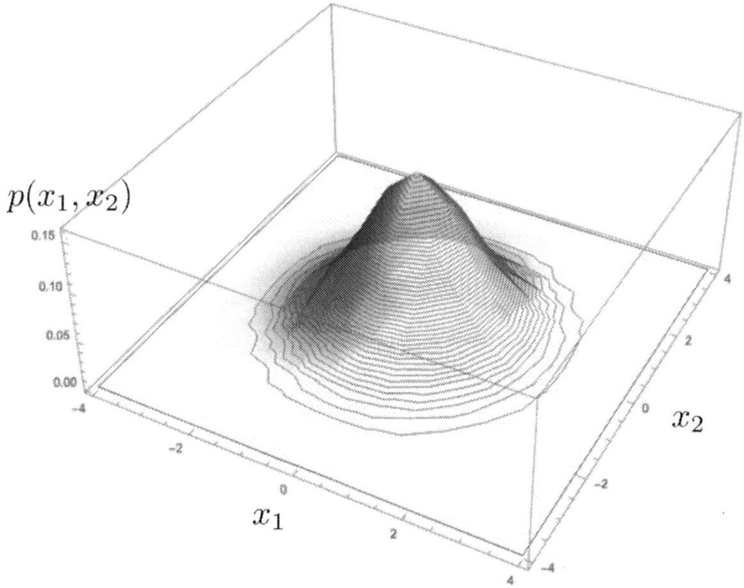

FIGURE A2.2 Probability density function with two random variables.

A2.1.3 White Noise

The autocorrelation function of the time series $z(k)$ is defined by

$$R(\tau) = E\left[z(k)z(k+\tau)\right], \tag{A2.16}$$

where τ denotes the time difference.

The time series $z(k)$ is called white noise when the autocorrelation function $R(\tau)$ becomes

$$R(\tau) = \begin{cases} \sigma_z^2 & (\tau = 0) \\ 0 & (\tau \neq 0) \end{cases}. \tag{A2.17}$$

The time series $z(k)$is uncorrelated when $\tau \neq 0$, because the autocorrelation function $R(\tau)$ is zero.

A2.2 Bayes' Theorem

The probability of event A occurring is defined as $P(A)$, the probability of event B occurring is $P(B)$ and the probability of events A and B occurring simultaneously is called the simultaneous probability and is defined as $P(A,B)$ (see Figure A2.3).

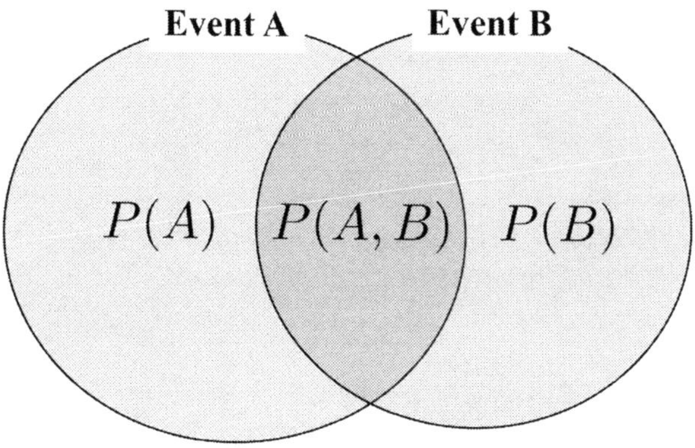

FIGURE A2.3 Events and probability.

The probability of event A occurring when event B occurs is described as $P(AB)$ and is called conditional probability and is defined as follows.

$$P(A/B) = \frac{P(A,B)}{P(B)}. \tag{A2.18}$$

If events A and B are independent, then we obtain

$$P(A,B) = P(A)P(B). \tag{A2.19}$$

The conditional probability is

$$P(A/B) = \frac{P(A)P(B)}{P(B)} = P(A), \tag{A2.20}$$

and is therefore not affected by event B.

Now, from Equation (A2.18), we obtain

$$P(A,B) = P(A|B)P(B), \tag{A2.21}$$

$$P(B,A) = P(B|A)P(A). \tag{A2.22}$$

Because $P(A,B) = P(B,A)$ by definition, we have

$$P(B|A) = \frac{P(A|B)P(B)}{P(A)}. \tag{A2.23}$$

Equation (A2.23) is called Bayes' theorem and is the fundamental theorem of Bayesian estimation.

If the random variables $\boldsymbol{x}$ and $\boldsymbol{y}$ are continuous vectors, Bayes' theorem can be expressed as follows:

$$p(\boldsymbol{x}|\boldsymbol{y}) = \frac{p(\boldsymbol{y}|\boldsymbol{x})p(\boldsymbol{x})}{\int p(\boldsymbol{y}|\boldsymbol{x})p(\boldsymbol{x})d\boldsymbol{x}} \propto p(\boldsymbol{y}|\boldsymbol{x})p(\boldsymbol{x}). \tag{A2.24}$$

A2.3 Least-Squares Estimation Method

Consider the problem of estimating the state (here assumed to be constant) from noisy sensor measurements (see Figure A2.4).

Assume that there is a linear relationship between the states $x(1), \cdots, x(n)$ and the measured value $z(i)$. In this case, the measurement is carried out m times to obtain the measurement value. It is assumed that each measurement contains random noise with different values:

$$\begin{aligned} z(1) &= H_{11}x(1) + \cdots + H_{1n}x(n) + v(1) \\ &\vdots \\ z(m) &= H_{m1}x(1) + \cdots + H_{mn}x(n) + v(m) \end{aligned}. \tag{A2.25}$$

where $v(i)$ is the noise added to the measured values.

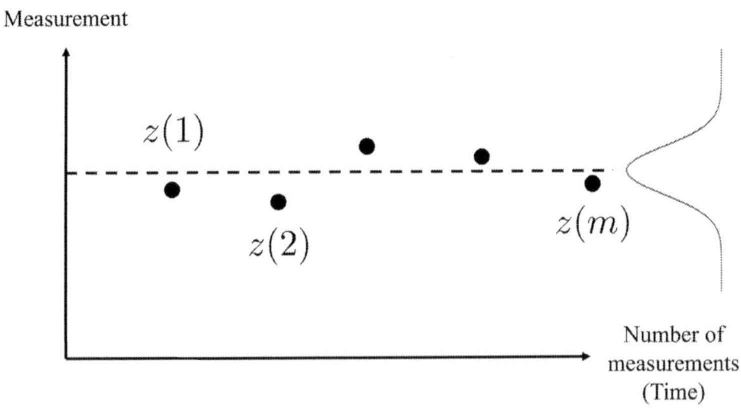

FIGURE A2.4 Least-squares estimation method.

Expressing Equation (A2.25) in matrix form gives

$$\underbrace{\begin{bmatrix} z(1) \\ \vdots \\ z(m) \end{bmatrix}}_{z} = \underbrace{\begin{bmatrix} H_{11} & \cdots & H_{1n} \\ \vdots & \ddots & \vdots \\ H_{m1} & \cdots & H_{mn} \end{bmatrix}}_{H} \underbrace{\begin{bmatrix} x(1) \\ \vdots \\ x(n) \end{bmatrix}}_{x} + \underbrace{\begin{bmatrix} v(1) \\ \vdots \\ v(m) \end{bmatrix}}_{v}.$$

$$z = \boldsymbol{Hx} + \boldsymbol{v}. \tag{A2.26}$$

Now we consider how to find the best estimate $\hat{\boldsymbol{x}}$ of state $\boldsymbol{x}$. The error between the observed value $\hat{z}$ (= $\boldsymbol{H}\hat{\boldsymbol{x}}$) predicted from the best estimate $\hat{\boldsymbol{x}}$ and the actual measured value z is

$$\tilde{z} = z - H\hat{\boldsymbol{x}} = [\tilde{z}(1), \cdots, \tilde{z}(m)]^T, \tag{A2.27}$$

where $\tilde{z}$ is called the residual.

To determine the best estimate $\hat{\boldsymbol{x}}$ of state $\boldsymbol{x}$ so that the cost function, which is the sum of the squares of the error $\tilde{z}(i)$ (i=1, $\cdots$, m) between the predicted and the actual observed values,

$$J = \tilde{z}(1)^2 + \cdots + \tilde{z}(m)^2, \tag{A2.28}$$

is minimized.

Equation (A2.28) can be rewritten as

$$J = \tilde{z}^T \tilde{z}$$

$$= (z - \boldsymbol{Hx})^T \left(z - \boldsymbol{Hx}\right)$$

$$= z^T z - \hat{\boldsymbol{x}}^T \boldsymbol{H}^T z - z^T H\hat{\boldsymbol{x}} + \hat{\boldsymbol{x}}^T \boldsymbol{H}^T H\hat{\boldsymbol{x}}. \tag{A2.29}$$

Differentiating Equation (A2.29) by $\hat{\boldsymbol{x}}$ and equating it to zero yields the best estimate $\hat{\boldsymbol{x}}$ as follows:

$$\frac{\partial J}{\partial \hat{x}} = -z^T \boldsymbol{H} - z^T \boldsymbol{H} + 2\hat{\boldsymbol{x}}^T \boldsymbol{H}^T \boldsymbol{H} = 0, \tag{A2.30}$$

$$\hat{\boldsymbol{x}} = \left(\boldsymbol{H}^T \boldsymbol{H}\right)^{-1} \boldsymbol{H}^T z. \tag{A2.31}$$

This method is called the least-squares estimation method.

Example 1.2

The true value of x is measured m times, and the sensor measurements $z(1)$, $z(2)$,⋯, $z(m)$ are obtained. It can be assumed that random sensor noises $v(1)$, $v(2)$,⋯, $v(m)$ are added to each measured value.

$$\begin{aligned} z(1) &= x + v(1) \\ &\vdots \\ z(m) &= x + v(m) \end{aligned}.$$

Using matrix notation, we obtain

$$\underbrace{\begin{bmatrix} z(1) \\ \vdots \\ z(m) \end{bmatrix}}_{z} = \underbrace{\begin{bmatrix} 1 \\ \vdots \\ 1 \end{bmatrix}}_{H} x + \underbrace{\begin{bmatrix} v(1) \\ \vdots \\ v(m) \end{bmatrix}}_{v}.$$

The best estimate $\hat{x}$ is obtained as

$$\begin{aligned} \hat{x} &= \left(\boldsymbol{H}^T \boldsymbol{H}\right)^{-1} \boldsymbol{H}^T z \\ &= \left(\begin{bmatrix} 1 & \cdots & 1 \end{bmatrix} \begin{bmatrix} 1 \\ \vdots \\ 1 \end{bmatrix}\right)^{-1} [1 \cdots 1] \begin{bmatrix} z(1) \\ \vdots \\ z(m) \end{bmatrix} \\ &= \frac{1}{m}\left[z(1) + \cdots + z(m)\right]. \end{aligned}$$

In this case, the average of the measured values is the best estimate.

A2.4 Weighted Least-Squares Estimation Method

Next, consider the case where the variance $\sigma(i)^2$ values for the noise in the sensor measurements $z(i)$ differ from each other (see Figure A2.5).

For example, when the accuracy of the measurements is high, the noise variance $\sigma(i)^2$ is small, and when the accuracy of the measurements is low, the noise variance $\sigma(i)^2$ is large. Therefore, consider the estimation to give less weight to measurements with low accuracy and more weight to those with high accuracy.

To perform such an estimation, we consider the following evaluation function, which weights the variance of the noise in Equation (A2.28):

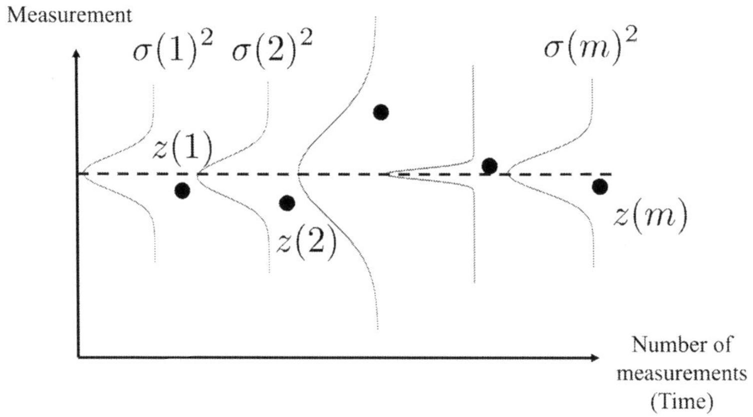

FIGURE A2.5 Weighted least squares estimation method.

$$J = \frac{\tilde{z}(1)^2}{\sigma(1)^2} + \cdots + \frac{\tilde{z}(m)^2}{\sigma(m)^2}. \tag{A2.32}$$

In this way, less accurate measurements have a lower weight for the cost function, thus improving the accuracy of the estimation.

By rewriting Equation (A2.32) using the following equation,

$$\boldsymbol{R} = \begin{bmatrix} \sigma(1)^2 & \cdots & 0 \\ \vdots & \ddots & \vdots \\ 0 & \cdots & \sigma(m)^2 \end{bmatrix} = \mathrm{diag}\left(\sigma(1)^2, \cdots, \sigma(m)^2\right), \tag{A2.33}$$

the cost function J to be minimized can be expressed as

$$\begin{aligned} J &= \tilde{\boldsymbol{z}}^T \boldsymbol{R}^{-1} \tilde{\boldsymbol{z}} \\ &= (\boldsymbol{z} - \boldsymbol{H}\hat{\boldsymbol{x}})^T \boldsymbol{R}^{-1} (\boldsymbol{z} - \boldsymbol{H}\hat{\boldsymbol{x}}) \\ &= \boldsymbol{z}^T \boldsymbol{R}^{-1} \boldsymbol{z} - \hat{\boldsymbol{x}}^T \boldsymbol{H}^T \boldsymbol{R}^{-1} \boldsymbol{z} - \boldsymbol{z}^T \boldsymbol{R}^{-1} \boldsymbol{H}\hat{\boldsymbol{x}} + \hat{\boldsymbol{x}}^T \boldsymbol{H}^T \boldsymbol{R}^{-1} \boldsymbol{H}\boldsymbol{x} \end{aligned} \tag{A2.34}$$

As in the previous section, by differentiating Equation (A2.34) by $\hat{\boldsymbol{x}}$ and equating it to zero, obtaining the best estimate $\hat{\boldsymbol{x}}$.

$$\frac{\partial J}{\partial \hat{\boldsymbol{x}}} = -\boldsymbol{z}^T \boldsymbol{R}^{-1} \boldsymbol{H} + \hat{\boldsymbol{x}}^T \boldsymbol{H}^T \boldsymbol{R}^{-1} \boldsymbol{H} = 0, \tag{A2.35}$$

$$\hat{x} = \left(H^T R^{-1} H\right)^{-1} H^T R^{-1} z. \tag{A2.36}$$

This method is called the weighted least-squares estimation method.

Example 1.3

As in Example 1.2, the true value of x is measured m times to obtain the sensor measurement values $z(1), z(2), \ldots, z(m)$. It can be assumed that each measurement is accompanied by a noise $v(1), v(2), \ldots, v(m)$ with different variance.

$$\underbrace{\begin{bmatrix} z(1) \\ \vdots \\ z(m) \end{bmatrix}}_{z} = \underbrace{\begin{bmatrix} 1 \\ \vdots \\ 1 \end{bmatrix}}_{H} x + \underbrace{\begin{bmatrix} v(1) \\ \vdots \\ v(m) \end{bmatrix}}_{v},$$

where

$$E\left[v(i)^2\right] = \sigma(i)^2 \left(i = 1, \cdots, m\right),$$

$$R = \begin{bmatrix} \sigma(1)^2 & \cdots & 0 \\ \vdots & \ddots & \vdots \\ 0 & \cdots & \sigma(m)^2 \end{bmatrix}.$$

The best estimate $\hat{x}$ is obtained from Equation (*A*2.36) as follows:

$$\begin{aligned} \hat{x} &= \left(H^T R^{-1} H\right)^{-1} H^T R^{-1} z \\ &= \left(\sum_{i=1}^{m} \frac{1}{\sigma(i)^2}\right)^{-1} \left(\frac{z(1)}{\sigma(1)^2} + \cdots + \frac{z(m)}{\sigma(m)^2}\right). \end{aligned}$$

In the above equation, if the variance of all noise is the same, the result is the same as that in Example 1.2.

A2.5 Sequential Least-Squares Estimation Method

In the estimation using Equation (A2.36), it is necessary to recalculate Equation (A2.36) repeatedly for each new measurement, which is very inefficient when the number of measurements increases. If the method could be changed to a sequential type, in which the estimates obtained until then are updated for each new measurement, this would improve the efficiency of the calculations. This method is called the sequential least-squares estimation method (see Figure A2.6).

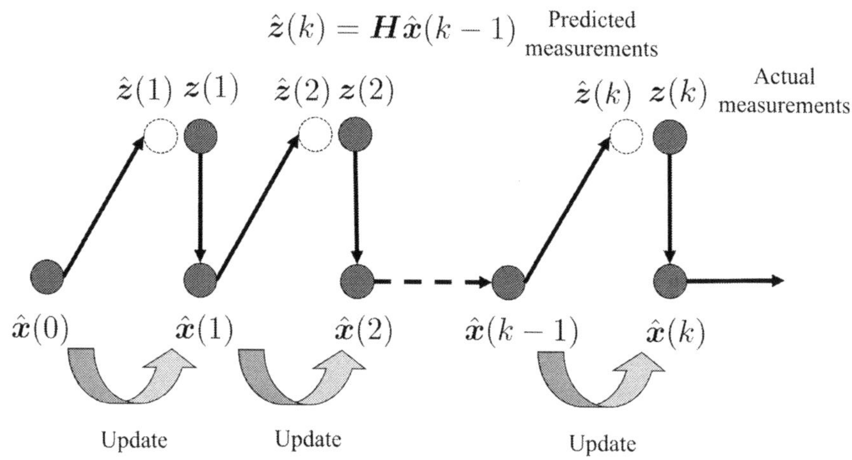

FIGURE A2.6 Sequential least-squares estimation method.

Consider modifying $\hat{x}(k-1)$ at time $k-1$ using the difference between the observed value $H\hat{x}(k-1)$ predicted from the state estimate $\hat{x}(k-1)$ and the value actually observed at time k, $z(k)$.

$$z(k) = Hx(k) + v(k), \tag{A2.37}$$

$$\hat{x}(k) = \hat{x}(k-1) + K(k)\left[z(k) - H\hat{x}(k-1)\right], \tag{A2.38}$$

where $K(k)$ is a coefficient matrix called the gain. The measurement noise $v(k)$ is white Gaussian noise with a mean of zero and a covariance matrix R.

Denote the true value as x and the difference between the true value and the estimated value $\hat{x}(k)$ (the estimation error) as $\tilde{x}(k)\left[= x - \hat{x}(k)\right)]$. The average value of the estimation error $\tilde{x}(k)$ can be calculated by the following equation:

$$\begin{aligned}
E\left[\tilde{x}(k)\right] &= E\left[x - \hat{x}(k)\right] \\
&= E\left[x - \hat{x}(k-1) - K(k)\left[z(k) - H\hat{x}(k-1)\right]\right] \\
&= E\left[\tilde{x}(k-1) - K(k)\left[Hx + v(k) - H\hat{x}(k-1)\right]\right] \\
&= E\left[\tilde{x}(k-1) - K(k)H\left[x - \hat{x}(k-1)\right] - K(k)v(k)\right] \\
&= E\left[\tilde{x}(k-1) - K(k)H\tilde{x}(k-1) - K(k)v(k)\right]
\end{aligned}$$

$$= \left[\boldsymbol{I} - \boldsymbol{K}(k)\boldsymbol{H} \right] E\left[\tilde{\boldsymbol{x}}(k-1) \right] - \boldsymbol{K}(k) E\left[\boldsymbol{v}(k) \right]$$

$$= \left[\boldsymbol{I} - \boldsymbol{K}(k)\boldsymbol{H} \right] E\left[\tilde{\boldsymbol{x}}(k-1) \right]. \tag{A2.39}$$

The estimation error covariance matrix can also be calculated by

$$\boldsymbol{P}(k) = E\left[\tilde{\boldsymbol{x}}(k) \tilde{\boldsymbol{x}}(k)^T \right]$$

$$= E\left[\left[\left(\boldsymbol{I} - \boldsymbol{K}(k)\boldsymbol{H} \right) \tilde{\boldsymbol{x}}(k-1) - \boldsymbol{K}(k)\boldsymbol{v}(k) \right] [\cdots]^T \right]$$

$$= \left[\boldsymbol{I} - \boldsymbol{K}(k)\boldsymbol{H} \right] E\left[\tilde{\boldsymbol{x}}(k-1) \tilde{\boldsymbol{x}}(k-1)^T \right] [\boldsymbol{I} - \boldsymbol{K}(k)\boldsymbol{H}]^T$$

$$- \boldsymbol{K}(k) E\left[\boldsymbol{v}(k) \tilde{\boldsymbol{x}}(k-1)^T \right] [\boldsymbol{I} - \boldsymbol{K}(k)\boldsymbol{H}]^T$$

$$- \left[\boldsymbol{I} - \boldsymbol{K}(k)\boldsymbol{H} \right] E\left[\tilde{\boldsymbol{x}}(k) \boldsymbol{v}(k)^T \right] \boldsymbol{K}(k)^T + \boldsymbol{K}(k) E\left[\boldsymbol{v}(k) \boldsymbol{v}(k)^T \right] \boldsymbol{K}(k)^T, \tag{A2.40}$$

where the estimation error and the noise are uncorrelated.

Note that because $E\left[\boldsymbol{v}(k) \tilde{\boldsymbol{x}}(k-1)^T \right] = 0, E\left[\tilde{\boldsymbol{x}}(k) \boldsymbol{v}(k)^T \right] = 0$, we obtain

$$\boldsymbol{P}(k) = \left[\boldsymbol{I} - \boldsymbol{K}(k)\boldsymbol{H} \right] \boldsymbol{P}(k-1) [\boldsymbol{I} - \boldsymbol{K}(k)\boldsymbol{H}]^T + \boldsymbol{K}(k)\boldsymbol{R}\boldsymbol{K}(k)^T. \tag{A2.41}$$

The cost function $J(k)$ that minimises the estimation error can be expressed as

$$J(k) = E\left[\left(x_1(k) - \hat{x}_1(k) \right)^2 + \cdots + \left(x_n(k) - \hat{x}_n(k) \right)^2 \right]$$

$$= E\left[\tilde{x}_1(k)^2 + \cdots + \tilde{x}_n(k)^2 \right]$$

$$= E\left[\tilde{\boldsymbol{x}}(k)^T \tilde{\boldsymbol{x}}(k) \right]$$

$$= E\left[\mathrm{Tr}\left(\tilde{\boldsymbol{x}}(k) \tilde{\boldsymbol{x}}(k)^T \right) \right] = \mathrm{Tr} E\left[\tilde{\boldsymbol{x}}(k) \tilde{\boldsymbol{x}}(k)^T \right] = \mathrm{Tr} \boldsymbol{P}(k). \tag{A2.42}$$

By differentiating the cost function $J(k)$ by $\boldsymbol{K}(k)$ and setting it to zero, we obtain[1]

$$\frac{\partial J(k)}{\partial \boldsymbol{K}(k)} = -2\left[\boldsymbol{I} - \boldsymbol{K}(k)\boldsymbol{H} \right] \boldsymbol{P}(k-1)\boldsymbol{H}^T + 2\boldsymbol{K}(k)\boldsymbol{R} = 0. \tag{A2.43}$$

From Equation (A2.43), the gain $\boldsymbol{K}(k)$ is obtained as

$$\boldsymbol{K}(k) = \boldsymbol{P}(k-1)\boldsymbol{H}^T\boldsymbol{S}(k-1)^{-1}, \tag{A2.44}$$

where

$$\boldsymbol{S}(k-1) = \boldsymbol{HP}(k-1)\boldsymbol{H}^T + \boldsymbol{R}\,. \tag{A2.45}$$

Now, rewriting Equation (A2.41) using Equation (A2.44), we obtain[2]

$$\boldsymbol{P}(k) = \left[\boldsymbol{I} - \boldsymbol{K}(k)\boldsymbol{H}\right]\boldsymbol{P}(k-1)\,. \tag{A2.46}$$

From the above, obtain the sequential least-squares estimation algorithm shown in Figure A2.7.

Step 1: Set the initial values as

$$\begin{aligned} \hat{\boldsymbol{x}}(0) &= E\left[\boldsymbol{x}(0)\right], \\ \boldsymbol{P}(0) &= E\left[\left[\boldsymbol{x}(0) - \hat{\boldsymbol{x}}(0)\right][\boldsymbol{x}(0) - \hat{\boldsymbol{x}}(0)]^T\right]. \end{aligned} \tag{A2.47}$$

Step 2: Obtain the measurement value $z(k)$.

$$z(k) = \boldsymbol{Hx}(k) + \boldsymbol{v}(k)\,, \tag{A2.48}$$

where the measurement noise $\boldsymbol{v}(k)$ is white Gaussian with a zero mean and a covariance matrix $\boldsymbol{R}$.

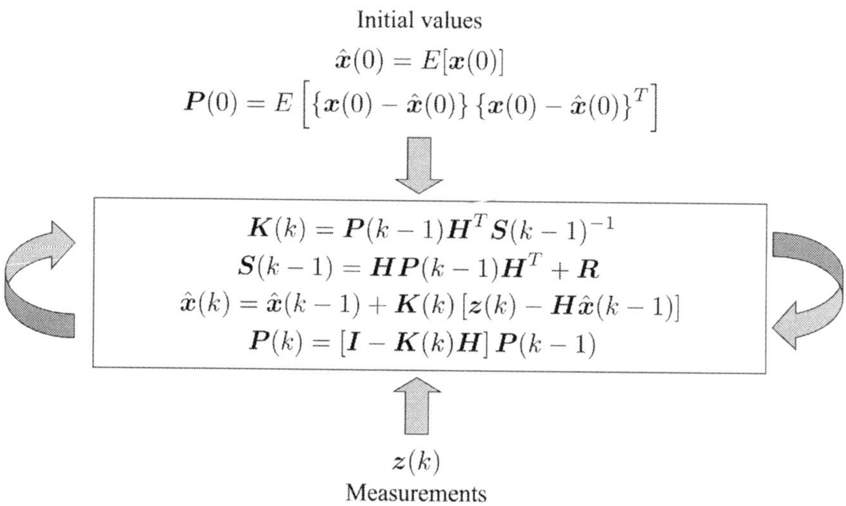

FIGURE A2.7 Sequential least-squares estimation algorithm.

Step 3: Update the estimates as

$$\boldsymbol{K}(k)=\boldsymbol{P}(k-1)\boldsymbol{H}^T\boldsymbol{S}(k-1)^{-1}, \tag{A2.49}$$

$$\boldsymbol{S}(k-1)=\boldsymbol{H}\boldsymbol{P}(k-1)\boldsymbol{H}^T+\boldsymbol{R}, \tag{A2.50}$$

$$\hat{\boldsymbol{x}}(k)=\hat{\boldsymbol{x}}(k-1)+\boldsymbol{K}(k)\left[z(k)-\boldsymbol{H}\hat{\boldsymbol{x}}(k-1)\right], \tag{A2.51}$$

$$\boldsymbol{P}(k)=\left[\boldsymbol{I}-\boldsymbol{K}(k)\boldsymbol{H}\right]\boldsymbol{P}(k-1). \tag{A2.52}$$

Step 4: Repeat Steps 2 and 3 for k=1, 2, ⋯, m.

Example 1.4

For the problem shown in Example 1.2, we construct a sequential least-squares estimation algorithm. The sensor measurements can be expressed by the following equation:

$$z(k)=Hx(k)+v(k)$$

where

$$H=1\,,$$
$$R=E\left[v(k)^2\right].$$

The initial value is given by Equation (A2.47), which gives

$$\begin{aligned}\hat{x}(0)&=E\left[x(0)\right],\\ P(0)&=E\left[\left[x(0)-\hat{x}(0)\right][x(0)-\hat{x}(0)]^T\right]\\ &=E\left[[x(0)-\hat{x}(0)]^2\right].\end{aligned}$$

Note that because

$$\begin{aligned}K(k)&=P(k-1)H^T\left[HP(k-1)H^T+R\right]^{-1}\\ &=\frac{P(k-1)}{P(k-1)+R},\\ P(k)&=\left[I-K(k)H\right]P(k-1)\\ &=\left(1-K(k)\right)P(k-1),\end{aligned}$$

we obtain

$$K(k)=\frac{P(0)}{kP(0)+R}.$$

The estimate $\hat{x}(k)$ can be calculated as

$$\begin{aligned}\hat{x}(k)&=\hat{x}(k-1)+K(k)[z(k)-H\hat{x}(k-1)\\&=\hat{x}(k-1)+K(k)\left[z(k)-H\hat{x}(k-1)\right]\\&=\frac{(k-1)P(0)+R}{kP(0)+R}\hat{x}(k-1)+\frac{P(0)}{kP(0)+R}z(k).\end{aligned}$$

When x is known, $P(0)=0$ and the gain $K(k)=0$. This shows that the best estimate is $\hat{x}=x(0)$, because no correction is made by the measurements.

However, if x is unknown, then $P(0)\to\infty$ and the best estimate can be calculated as

$$\begin{aligned}\hat{x}(k)&=\frac{(k-1)P(0)}{kP(0)}\hat{x}(k-1)+\frac{P(0)}{kP(0)}z(k)\\&=\frac{(k-1)}{k}\hat{x}(k-1)+\frac{1}{k}z(k)\\&=\frac{1}{k}\left[(k-1)\hat{x}(k-1)+z(k)\right].\end{aligned}$$

If we write the above equation in concrete form, we obtain

$$\begin{aligned}\hat{x}(1)&=z(1)\\\hat{x}(2)&=\frac{1}{2}\left[\hat{x}(1)+z(2)\right]=\frac{1}{2}\left[z(1)+z(2)\right]\\&\vdots\\\hat{x}(k)&=\frac{1}{2}\left[\hat{x}(k-1)+z(k)\right]=\frac{1}{k}\left[z(1)+z(2)+\cdots+z(k)\right].\end{aligned}$$

The results show that the average of the observations is the best estimate, which is consistent with the results in Example 1.2.

A2.6 Linear Kalman Filter

In this section, the sequential least-squares method derived in the previous section is extended to the KF.

The discrete-time state-space representation of the system can be expressed as

$$x(k) = Fx(k-1) + Gu(k-1) + w(k-1), \tag{A2.53}$$

$$z(k) = Hx(k) + v(k), \tag{A2.54}$$

where $x(k)$, $u(k)$ and $z(k)$ are the state, input and sensor measurements, respectively; $w(k)$ is the system noise (also called plant noise), which is white Gaussian noise with a zero mean and a covariance matrix Q, and $v(k)$ is the sensor noise, with zero mean and white Gaussian noise with a covariance matrix R.

We define a prior estimate and a posteriori estimate of the state $x(k)$ at time k. The prior estimate is an estimate (prediction) based on the measured values $z(1), \cdots, z(k-1)$ up to time $k-1$, and is described as follows:

$$\hat{x}(k/k-1) = E[(x(k)/z(1), \cdots, z(k-1))]. \tag{A2.55}$$

The posterior estimate is based on the measured values $z(1), \cdots, z(k)$ up to time k and is written as

$$\hat{x}(k/k) = E[(x(k)/z(1), \cdots, z(k))]. \tag{A2.56}$$

This is the estimate to be obtained at time k.

Now consider extending the sequential least-squares estimation algorithm to work with Equations (A2.55) and (A2.56) and their covariance matrices. In the sequential least-squares estimation algorithm, the state estimate and covariance matrix are rewritten before and after the measurement value $z(k)$ as follows (see Figure A2.8):

$$\begin{gathered}\hat{x}(k-1) \to \hat{x}(k|k-1), \\ P(k-1) \to P(k|k-1), \\ \hat{x}(k) \to \hat{x}(k|k), \\ P(k) \to P(k|k).\end{gathered}$$

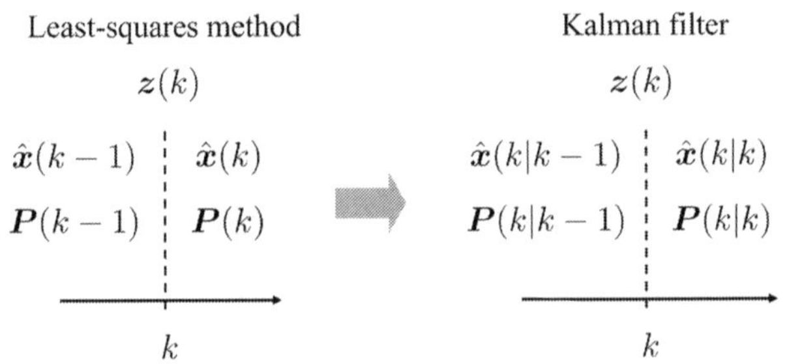

FIGURE A2.8 Sequential least-squares and Kalman Filter.

This allows the time update, which predicts the state one step after the measurement $z(k-1)$ and the state equation, and the measurement update, which modifies the estimate from the latest observation $z(k)$, as shown in Figure A2.9.

The structure of the sequential estimation by the KF is shown in Figure A2.10. The time and measurement update algorithms are summarised below.

- Time update algorithm

The prediction is formulated as

$$\hat{\boldsymbol{x}}(k|k-1) = \boldsymbol{F}\hat{\boldsymbol{x}}(k-1|k-1) + \boldsymbol{G}\boldsymbol{u}(k-1). \tag{A2.57}$$

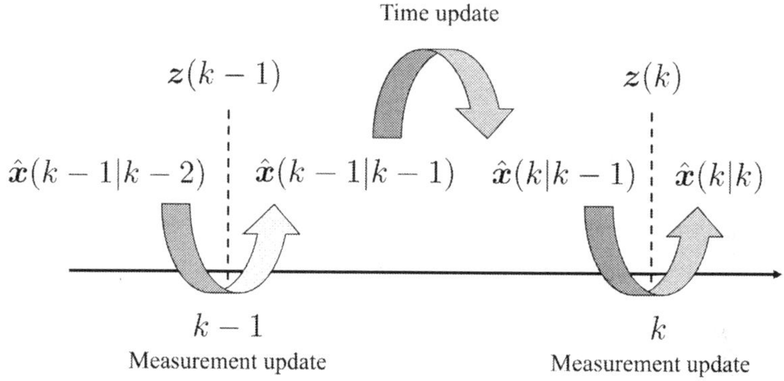

FIGURE A2.9 Time update and measurement update.

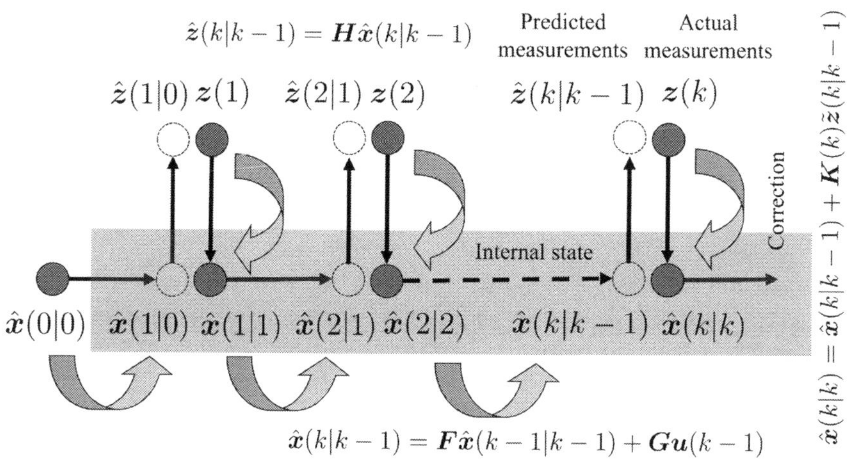

FIGURE A2.10 Sequential estimation by Kalman filter.

The prediction error covariance matrix is

$$\boldsymbol{P}(k|k-1) = \boldsymbol{F}\boldsymbol{P}(k-1|k-1)\boldsymbol{F}^T + \boldsymbol{Q}. \tag{A2.58}$$

- Measurement update algorithm

The Kalman gain is

$$\boldsymbol{K}(k) = \boldsymbol{P}(k|k-1)\boldsymbol{H}^T\boldsymbol{S}(k|k-1)^{-1}. \tag{A2.59}$$

The measurement prediction error covariance matrix is given by

$$\boldsymbol{S}(k|k-1) = \boldsymbol{H}\boldsymbol{P}(k|k-1)\boldsymbol{H}^T + \boldsymbol{R}. \tag{A2.60}$$

The predicted measured values (measurement predictions) are obtained as follows:

$$\hat{\boldsymbol{z}}(k|k-1) = \boldsymbol{H}\hat{\boldsymbol{x}}(k|k-1). \tag{A2.61}$$

The estimation equation is

$$\begin{aligned}\hat{\boldsymbol{x}}(k|k) &= \hat{\boldsymbol{x}}(k|k-1) + \boldsymbol{K}(k)\left[\boldsymbol{z}(k) - \hat{\boldsymbol{z}}(k|k-1)\right]\\ &= \hat{\boldsymbol{x}}(k|k-1) + \boldsymbol{K}(k)\left[\boldsymbol{z}(k) - \boldsymbol{H}\hat{\boldsymbol{x}}(k|k-1)\right].\end{aligned} \tag{A2.62}$$

The estimation error covariance matrix is given by[3]

$$\boldsymbol{P}(k|k) = \left[\boldsymbol{I} - \boldsymbol{K}(k)\boldsymbol{H}\right]\boldsymbol{P}(k|k-1). \tag{A2.63}$$

The algorithm for the KF is summarised in Figure A2.11.

A2.7 Extended Kalman Filter

The previous section describes the KF when the equations for the state and observation are linear. This section describes the extended KF that is applied when they are non-linear. The extended KF is historically old and easy to understand. The non-linear state equation and measurement equation for a discrete-time system are described in the following equations:

$$\boldsymbol{x}(k) = \boldsymbol{f}\left[\boldsymbol{x}(k-1), \boldsymbol{u}(k-1), \boldsymbol{w}(k-1)\right], \tag{A2.64}$$

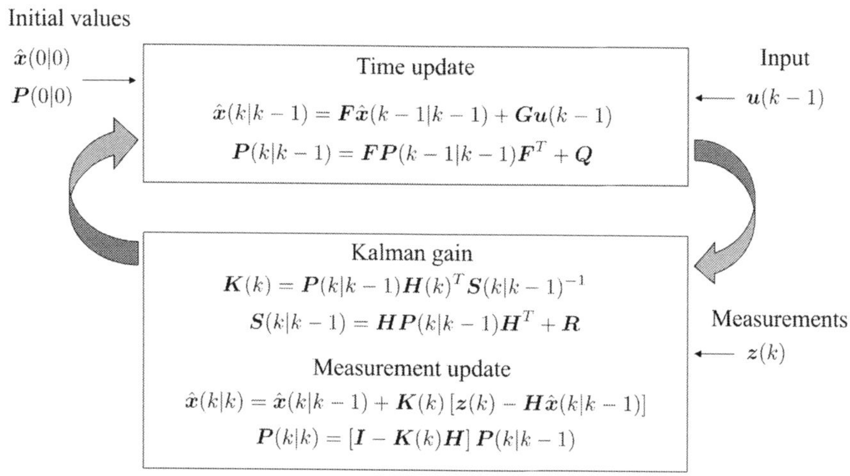

FIGURE A2.11 Kalman filter algorithm.

$$z(k) = h[x(k), v(k)], \tag{A2.65}$$

where $x(k)$, $u(k)$ and $z(k)$ are the state, input and sensor measurements, respectively; $w(k-1)$ is the system noise, which is white Gaussian noise with a zero mean and a covariance matrix Q; $v(k)$ is the sensor noise with a zero mean and white Gaussian noise with a covariance matrix R, and $f[\cdot]$ and $h[\cdot]$ represent non-linear functions.

With a Taylor expansion around the estimates $x(k-1) = \hat{x}(k-1|k-1)$ and $w(k-1) = 0$ in Equation (A2.64), and terms above the second-order term are assumed to be negligibly small, we obtain the linear equation of state as

$$\begin{aligned}
x(k) &= f[\hat{x}(k-1|k-1), u(k-1), 0] \\
&\quad + \left[\frac{\partial f}{\partial w}\right]_{x(k-1)=\hat{x}(k|k-1)} [x(k-1) - \hat{x}(k-1 \mid k-1)] \\
&\quad + \left[\frac{\partial f}{\partial w}\right]_{w(k-1)=0} w(k-1) \\
&= f[\hat{x}(k-1 \mid k-1), u(k-1), 0] \\
&\quad + F(k-1)[x(k-1) - \hat{x}(k-1 \mid k-1)] + L(k-1)w(k-1) \\
&= F(k-1)x(k-1) + f[\hat{x}(k-1 \mid k-1), u(k-1), 0] \\
&\quad - F(k-1)\hat{x}(k-1 \mid k-1) + L(k-1)w(k-1) \\
&= F(k-1)x(k-1) + \tilde{u}(k-1) + \tilde{w}(k-1).
\end{aligned} \tag{A2.66}$$

The known signal $\tilde{u}(k-1)$ and the system noise $\tilde{w}(k-1)$ can be expressed as

$$\tilde{\boldsymbol{u}}(k-1)=\boldsymbol{f}\left[\hat{\boldsymbol{x}}(k-1|k-1),\boldsymbol{u}(k-1),0\right]-\boldsymbol{F}(k-1)\hat{\boldsymbol{x}}(k-1|k-1), \quad \text{(A2.67)}$$

$$\tilde{\boldsymbol{w}}(k-1)=\boldsymbol{L}(k-1)\boldsymbol{w}(k-1). \quad \text{(A2.68)}$$

where $\boldsymbol{F}(k-1)$ and $\boldsymbol{L}(\mathrm{k}-1)$ are Jacobian matrices.

For example, taking $\boldsymbol{F}(k-1)$ as an example, if the dimension of the state $\boldsymbol{x}(k)$ is n-dimensional, $\boldsymbol{F}(k-1)$ can be written as

$$\boldsymbol{F}(k-1)=\frac{\partial \boldsymbol{f}}{\partial \boldsymbol{x}(k-1)}=\begin{bmatrix} \frac{\partial f_1}{\partial x_1(k-1)} & \frac{\partial f_1}{\partial x_2(k-1)} & \cdots & \frac{\partial f_1}{\partial x_n(k-1)} \\ \vdots & \vdots & \cdots & \vdots \\ \frac{\partial f_n}{\partial x_1(k-1)} & \frac{\partial f_n}{\partial x_2(k-1)} & \cdots & \frac{\partial f_n}{\partial x_n(k-1)} \end{bmatrix}. \quad \text{(A2.69)}$$

Because $\boldsymbol{w}(k-1)$ has a zero mean and the covariance matrix is $\boldsymbol{Q}$, from Equation (A2.68), $\boldsymbol{w}(k-1)$ has a zero mean and the covariance matrix is $\boldsymbol{L}(k-1)\boldsymbol{Q}\boldsymbol{L}(k-1)^T$.

However, if the right-hand side of Equation (A2.65) is expanded using the Taylor method around the predicted values $\boldsymbol{x}(k)=\hat{\boldsymbol{x}}(k|k-1)$ and $\boldsymbol{v}(k)=0$, and omitting terms above the quadratic term, assuming that they are negregibly small, the following linear measurement equation is obtained:

$$\begin{aligned} \boldsymbol{z}(k)&=\boldsymbol{h}\left[\hat{\boldsymbol{x}}(k|k-1),0\right]+\left[\frac{\partial \boldsymbol{h}}{\partial \boldsymbol{x}}\right]_{\boldsymbol{x}(k)=\hat{\boldsymbol{x}}(k|k-1)}\left[\boldsymbol{x}(k)-\hat{\boldsymbol{x}}(k|k-1)\right] \\ &\quad+\left[\frac{\partial \boldsymbol{h}}{\partial \boldsymbol{v}}\right]_{\boldsymbol{v}(k)=0}\boldsymbol{v}(k) \\ &=\boldsymbol{h}\left[\hat{\boldsymbol{x}}(k|k-1),0\right]+\boldsymbol{H}(k)\left[\boldsymbol{x}(k)-\hat{\boldsymbol{x}}(k|k-1)\right]+\boldsymbol{M}(k)\boldsymbol{v}(k) \\ &=\boldsymbol{H}(k)\boldsymbol{x}(k)+\left[\boldsymbol{h}\left[\hat{\boldsymbol{x}}(k|k-1),0\right]-\boldsymbol{H}(k)\hat{\boldsymbol{x}}(k|k-1)\right]+\boldsymbol{M}(k)\boldsymbol{v}(k) \\ &=\boldsymbol{H}(k)\boldsymbol{x}(k)+\boldsymbol{y}(k)+\hat{\boldsymbol{v}}(k). \end{aligned} \quad \text{(A2.70)}$$

where the known signal $\boldsymbol{y}(k)$ and the sensor noise $\hat{\boldsymbol{v}}(k)$ can be expressed, respectively, as

$$\boldsymbol{y}(k)=\boldsymbol{h}\left[\hat{\boldsymbol{x}}(k|k-1),0\right]-\boldsymbol{H}(k)\hat{\boldsymbol{x}}(k|k-1)], \quad \text{(A2.71)}$$

$$\hat{\hat{v}}(k) = M(k)v(k). \tag{A2.72}$$

where $H(k)$ and $M(k)$ are Jacobian matrices.

Taking $H(k)$ as an example, if the dimension of the measured value $z(k)$ is m-dimensional and the dimension of the state $x(k)$ is n-dimensional, $H(k)$ can be written as

$$H(k) = \frac{\partial h}{\partial x(k)} = \begin{bmatrix} \frac{\partial h_1}{\partial x_1(k)} & \frac{\partial h_1}{\partial x_2(k)} & \cdots & \frac{\partial h_1}{\partial x_n(k)} \\ \vdots & \vdots & \cdots & \vdots \\ \frac{\partial h_m}{\partial x_1(k)} & \frac{\partial h_m}{\partial x_2(k)} & \cdots & \frac{\partial h_m}{\partial x_n(k)} \end{bmatrix}. \tag{A2.73}$$

Because $v(k)$ has a zero mean and the covariance matrix is R, from Equation (A2.72), $v(k)$ has a zero mean and the covariance matrix is $M(k)RM(k)^T$.

The linearized equation of state in Equation (A2.66) and the measurement equation in Equation (A2.70) can be summarized as

$$x(k) = F(k-1)x(k-1) + \hat{u}(k-1) + \hat{w}(k-1), \tag{A2.74}$$

$$z(k) = H(k)x(k) + y(k) + \hat{v}(k). \tag{A2.75}$$

By applying the linear KF described in Section A2.6 to the above equation, the time update algorithm and measurement update algorithm can be obtained as shown below.

- Time update algorithm

The prediction is made as follows:

$$\hat{x}(k|k-1) = f\left[\hat{x}(k-1|k-1), u(k-1), 0\right]. \tag{A2.76}$$

The prediction error covariance matrix is

$$P(k|k-1) = F(k-1)P(k-1|k-1)F(k-1)^T + L(k-1)QL(k-1)^T. \tag{A2.77}$$

- Measurement update algorithm

The Kalman gain is

$$\boldsymbol{K}(k) = \boldsymbol{P}(k|k-1)\boldsymbol{H}(k)^T \boldsymbol{S}(k|k-1)^{-1}. \tag{A2.78}$$

The measurement prediction error covariance matrix is given by

$$\boldsymbol{S}(k|k-1) = \boldsymbol{H}(k)\boldsymbol{P}(k|k-1)\boldsymbol{H}(k)^T + \boldsymbol{M}(k)\boldsymbol{R}\boldsymbol{M}(k)^T. \tag{A2.79}$$

The estimates are obtained as follows:

$$\hat{\boldsymbol{x}}(k|k) = \hat{\boldsymbol{x}}(k|k-1) + \boldsymbol{K}(k)\left[z(k) - \boldsymbol{h}\left[\hat{\boldsymbol{x}}(k|k-1), 0\right]\right]. \tag{A2.80}$$

The estimation error covariance matrix is

$$\boldsymbol{P}(k|k) = \left[\boldsymbol{I} - \boldsymbol{K}(k)\boldsymbol{H}(k)\right]\boldsymbol{P}(k|k-1). \tag{A2.81}$$

A3 CONTINUOUS-TIME KALMAN FILTER

The equations of state and measurement of a dynamic system in continuous time are described below.

$$\dot{\boldsymbol{x}}(t) = \boldsymbol{A}\boldsymbol{x}(t) + \boldsymbol{B}\boldsymbol{u}(t) + \tilde{\boldsymbol{w}}(t), \tag{A3.1}$$

$$z(t) = \boldsymbol{C}\boldsymbol{x}(t) + \tilde{\boldsymbol{v}}(t), \tag{A3.2}$$

where $\tilde{\boldsymbol{w}}(t)$ is white Gaussian noise with a mean of zero and a covariance matrix $\boldsymbol{W}(t)$ and $\tilde{\boldsymbol{v}}(t)$ is white Gaussian noise with a mean of zero and a covariance matrix $\boldsymbol{V}(t)$. It's assumed that $\tilde{\boldsymbol{w}}(t)$ and $\tilde{\boldsymbol{v}}(t)$ are uncorrelated.

Euler's approximation for Equations (A1.15) and (A1.19), when the sampling period τ is small, gives

$$\boldsymbol{x}(k) = (\boldsymbol{I} + \boldsymbol{A}\tau)\boldsymbol{x}(k-1) + \boldsymbol{B}\tau u(k-1) + \boldsymbol{w}(k-1), \tag{A3.3}$$

$$z(k) = \boldsymbol{C}\boldsymbol{x}(k) + \boldsymbol{v}(k), \tag{A3.4}$$

where $\boldsymbol{w}(k)$ is the system noise, which is white Gaussian noise with a zero mean and a covariance matrix $\boldsymbol{W}\tau$ and $\boldsymbol{v}(k)$ is the sensor noise, with a zero mean and white Gaussian with covariance matrix $\boldsymbol{V}/\tau$ (see Section A1.2).

Equations (A2.58), (A2.59) and (A2.63) become

$$\boldsymbol{P}(k|k-1) = (\boldsymbol{I}+\boldsymbol{A}\tau)\boldsymbol{P}(k-1|k-1)(\boldsymbol{I}+\boldsymbol{A}\tau)^T + \boldsymbol{W}\tau, \tag{A3.5}$$

$$\boldsymbol{K}(k) = \boldsymbol{P}(k|k-1)\boldsymbol{C}^T\left[\boldsymbol{C}\boldsymbol{P}(k|k-1)\boldsymbol{C}^T + \frac{\boldsymbol{V}}{\tau}\right]^{-1}, \tag{A3.6}$$

$$\boldsymbol{P}(k|k) = \left[\boldsymbol{I} - \boldsymbol{K}(k)\boldsymbol{C}\right]\boldsymbol{P}(k|k-1). \tag{A3.7}$$

From Equation (A3.6), we obtain

$$\frac{1}{\tau}\boldsymbol{K}(k) = \boldsymbol{P}(k|k-1)\boldsymbol{C}^T[\boldsymbol{C}\boldsymbol{P}(k|k-1)\boldsymbol{C}^T\tau + \boldsymbol{V}]^{-1}. \tag{A3.8}$$

When the limit is $\tau \to 0$, we obtain

$$\lim_{\tau\to 0}\frac{1}{\tau}\boldsymbol{K}(k) = \boldsymbol{P}(k|k-1)\boldsymbol{C}^T\boldsymbol{V}^{-1}. \tag{A3.9}$$

The prediction error covariance matrix can be expressed as

$$\begin{aligned}\boldsymbol{P}(k+1|k) &= (\boldsymbol{I}+\boldsymbol{A}\tau)\boldsymbol{P}(k|k)(\boldsymbol{I}+\boldsymbol{A}\tau)^T + \boldsymbol{W} \\ &= \boldsymbol{P}(k|k) + \left[\boldsymbol{A}\boldsymbol{P}(k|k) + \boldsymbol{P}(k|k)\boldsymbol{A}^T + \boldsymbol{W}\right]\tau.\end{aligned} \tag{A3.10}$$

Substituting Equation (A3.7) into Equation (A3.10), we obtain

$$\begin{aligned}&\boldsymbol{P}(k+1|k) = (\boldsymbol{I} - \boldsymbol{K}(k)\boldsymbol{C})\boldsymbol{P}(k|k-1) + \\ &\left[\boldsymbol{A}(\boldsymbol{I}-\boldsymbol{K}(k)\boldsymbol{C})\boldsymbol{P}(k|k-1) + (\boldsymbol{I}-\boldsymbol{K}(k)\boldsymbol{C})\boldsymbol{P}(k|k-1)\boldsymbol{A}^T + \boldsymbol{W}\right]\tau.\end{aligned} \tag{A3.11}$$

Thus,

$$\frac{\boldsymbol{P}(k+1|k) - \boldsymbol{P}(k|k-1)}{\tau}$$

$$= [\boldsymbol{AP}(k|k-1) + \boldsymbol{P}(k|k-1)\boldsymbol{A}^T + \boldsymbol{W} - \boldsymbol{AK}(k)\boldsymbol{CP}(k|k-1)$$

$$-\boldsymbol{K}(k)\boldsymbol{CP}(k|k-1)\boldsymbol{A}^T] - \frac{1}{\tau}\boldsymbol{K}(k)\boldsymbol{CP}(k|k-1). \quad \text{(A3.12)}$$

Prediction error covariance matrix $\boldsymbol{P}(t)$ satisfies $\boldsymbol{P}(k\tau) = \boldsymbol{P}(k|k-1)$. When the limit is $\tau \to 0$, we obtain

$$\dot{\boldsymbol{P}}(t) = \boldsymbol{AP}(t) + \boldsymbol{P}(t)\boldsymbol{A}^T - \boldsymbol{P}(t)\boldsymbol{C}^T\boldsymbol{V}^{-1}\boldsymbol{CP}(t) + \boldsymbol{W}. \quad \text{(A3.13)}$$

In the above formulation, Equation (A3.9) was applied. Equation (A3.13) is called a matrix Riccati differential equation.

The solution of the Equation (A3.13) tends to a bounded steady-state value if $\lim_{\tau \to 0} \boldsymbol{P}(t) = \boldsymbol{P}$ is bounded. In this case, because $\dot{\boldsymbol{P}}(t) \to 0$ for large t, we obtain the continuous algebraic Riccati equation.

$$\boldsymbol{AP} + \boldsymbol{PA}^T - \boldsymbol{PC}^T\boldsymbol{V}^{-1}\boldsymbol{CP} + \boldsymbol{W} = \boldsymbol{0}. \quad \text{(A3.14)}$$

The prediction equation can be written as

$$\hat{\boldsymbol{x}}(k|k-1) = (\boldsymbol{I} + \boldsymbol{A}\tau)\hat{\boldsymbol{x}}(k-1|k-1) + \boldsymbol{B}\tau\boldsymbol{u}(k-1). \quad \text{(A3.15)}$$

The estimation equation can be obtained using Equation (A3.15) as

$$\hat{\boldsymbol{x}}(k|k) = \hat{\boldsymbol{x}}(k|k-1) + \boldsymbol{K}(k)\left[\boldsymbol{z}(k) - \boldsymbol{C}\hat{\boldsymbol{x}}(k|k-1)\right]$$

$$= (\boldsymbol{I} + \boldsymbol{A}\tau)\hat{\boldsymbol{x}}(k-1|k-1) + \boldsymbol{B}\tau\boldsymbol{u}(k-1)$$

$$+\boldsymbol{K}(k)\left[\boldsymbol{z}(k) - \boldsymbol{C}(\boldsymbol{I} + \boldsymbol{A}\tau)\hat{\boldsymbol{x}}(k-1|k-1) - \boldsymbol{CB}\tau\boldsymbol{u}(k-1)\right]. \quad \text{(A3.16)}$$

Thus,

$$\frac{\hat{\boldsymbol{x}}(k|k) - \hat{\boldsymbol{x}}(k-1|k-1)}{\tau} = \boldsymbol{A}\hat{\boldsymbol{x}}(k-1|k-1) + \boldsymbol{Bu}(k-1)$$

$$+\frac{1}{\tau}\boldsymbol{K}(k)\left[\boldsymbol{z}(t) - \boldsymbol{C}\hat{\boldsymbol{x}}(k|k-1) - \boldsymbol{C}\left(\boldsymbol{A}\hat{\boldsymbol{x}}(k-1|k-1) + \boldsymbol{Bu}(k-1)\right)\tau\right]. \quad \text{(A3.17)}$$

In Equation (A3.17), $\hat{\boldsymbol{x}}(t)$ satisfies $\hat{\boldsymbol{x}}(k\tau) = \hat{\boldsymbol{x}}(k|k)$. In the limit as $\tau \to 0$, we obtain

$$\dot{\boldsymbol{x}}(t) = \boldsymbol{A}\hat{\boldsymbol{x}}(t) + \boldsymbol{Bu}(t) + \boldsymbol{P}(t)\boldsymbol{C}^T\boldsymbol{V}^{-1}\left[\boldsymbol{z}(t) - \boldsymbol{C}\hat{\boldsymbol{x}}(t)\right]. \quad \text{(A3.18)}$$

- Summary of continuous-time Kalman filter

The state and measurement equations are

$$\dot{\boldsymbol{x}}(t) = \boldsymbol{A}\boldsymbol{x}(t) + \boldsymbol{B}\boldsymbol{u}(t) + \tilde{\boldsymbol{w}}(t), \tag{A3.19}$$

$$z(t) = \boldsymbol{C}\boldsymbol{x}(t) + \tilde{\boldsymbol{w}}(t), \tag{A3.20}$$

where $\tilde{\boldsymbol{w}}(t)$ is the system noise, which is white Gaussian noise with a zero mean and a covariance matrix $\boldsymbol{W}$, and $\tilde{\boldsymbol{w}}(t)$is the sensor noise, with a zero mean and white Gaussian with covariance matrix $\boldsymbol{V}$. It's assumed that $\tilde{\boldsymbol{w}}(t)$ and $\tilde{\boldsymbol{v}}(t)$ are uncorrelated (i.e., $E\left[\tilde{\boldsymbol{w}}(\boldsymbol{t})\tilde{\boldsymbol{v}}(\boldsymbol{t})\right] = 0$).

The estimation error covariance is

$$\dot{\boldsymbol{P}}(t) = \boldsymbol{A}\boldsymbol{P}(t) + \boldsymbol{P}(t)\boldsymbol{A}^T - \boldsymbol{P}(t)\boldsymbol{C}^T\boldsymbol{V}^{-1}\boldsymbol{C}\boldsymbol{P}(t) + \boldsymbol{W}, \tag{A3.21}$$

or

$$\dot{\boldsymbol{P}}(t) = \boldsymbol{A}\boldsymbol{P}(t) + \boldsymbol{P}(t)\boldsymbol{A}^T - \boldsymbol{K}(t)\boldsymbol{R}\boldsymbol{K}(t)^T + \boldsymbol{W}. \tag{A3.22}$$

The Kalman gain is

$$\boldsymbol{K}(t) = \boldsymbol{P}(t)\boldsymbol{C}^T\boldsymbol{V}^{-1}. \tag{A3.23}$$

The estimation equation is

$$\dot{\boldsymbol{x}}(t) = \boldsymbol{A}\hat{\boldsymbol{x}}(t) + \boldsymbol{B}\boldsymbol{u}(t) + \boldsymbol{K}(t)\left[z(t) - \boldsymbol{C}\hat{\boldsymbol{x}}(t)\right]. \tag{A3.24}$$

Note that if $\tilde{\boldsymbol{w}}(t)$ and $\tilde{\boldsymbol{v}}(t)$ are correlated, $E\left[\tilde{\boldsymbol{w}}(t)\tilde{\boldsymbol{v}}(t)\right] = \boldsymbol{N}$, and the following Kalman gain and estimation error covariance equations are used:

$$\boldsymbol{K}(t) = (\boldsymbol{P}(t)\boldsymbol{C}^T + \boldsymbol{N})\boldsymbol{V}^{-1}, \tag{A3.25}$$

$$\dot{\boldsymbol{P}}(t) = \boldsymbol{A}\boldsymbol{P}(t) + \boldsymbol{P}(t)\boldsymbol{A}^T + \boldsymbol{W} - \left(\boldsymbol{P}(t)\boldsymbol{C}^T + \boldsymbol{N}\right)\boldsymbol{V}^{-1}\left(\boldsymbol{P}(t)\boldsymbol{C}^T + \boldsymbol{N}\right)^T. \tag{A3.26}$$

$\dot{\boldsymbol{P}}(t) = 0$ if the solution of the Riccati Equation $(A3.21)$ tends to a bounded steady-state value $\boldsymbol{P}$ for large t. In this case, we can obtain the steady-state continuous-time KF.

- Steady-state continuous-time Kalman filter

The estimation error covariance is

$$\boldsymbol{AP} + \boldsymbol{PA}^T - \boldsymbol{PC}^T\boldsymbol{V}^{-1}\boldsymbol{CP} + \boldsymbol{W} = \boldsymbol{0}, \tag{A3.27}$$

The Kalman gain is

$$\boldsymbol{K} = \boldsymbol{PC}^T\boldsymbol{V}^{-1}. \tag{A3.28}$$

The estimation equation is

$$\dot{\boldsymbol{x}}(t) = \boldsymbol{A}\hat{\boldsymbol{x}}(t) + \boldsymbol{Bu}(t) + \boldsymbol{K}\left[\boldsymbol{z}(t) - \boldsymbol{C}\hat{\boldsymbol{x}}(t)\right]. \tag{A3.29}$$

A4 STATE-SPACE MODELLING OF VEHICLE DYNAMICS USING THE NEWMARK β METHOD

Time-series analysis problems, such as vehicle dynamics simulation, require time integration of the equations of motion. This method is based on differential equations and is used to obtain the time history response of the vehicle through a large number of iterations. Common numerical integration methods include the Euler and Runge–Kutta methods.

However, these methods do not include acceleration as a state quantity in the first-order differential equations. This means that the velocity must be differentiated to obtain the car body acceleration when track irregularity is given as an input.

Therefore, it is better to use the Newmark β method of second-order differential equations when the vehicle model is represented by MCK form equations and acceleration is treated as a state quantity. The Newmark β method also has a damping effect that removes high-frequency components, thus preventing divergence of the solution.

The Newmark β method for second-order differential forms is therefore used as a time integration method. The acceleration $\boldsymbol{Z}$, velocity $\dot{\boldsymbol{Z}}$ and displacement $\boldsymbol{Z}$ are considered to divide the time into small time periods τ as shown in Figure A4.1. The equation of motion at time k is expressed as

$$\boldsymbol{M}\ddot{\boldsymbol{Z}}(k) + \boldsymbol{C}\dot{\boldsymbol{Z}}(k) + \boldsymbol{K}(k) = \boldsymbol{D}\dot{\boldsymbol{r}}(k) + \boldsymbol{E}\boldsymbol{r}(k). \tag{A4.1}$$

If the acceleration is assumed to vary linearly from time k to $k+1$, the acceleration $\boldsymbol{Z}_{\eta}$ at time η is

$$\ddot{\boldsymbol{Z}}_{\eta} = \ddot{\boldsymbol{Z}}(k) + \frac{\eta}{\tau}\left[\ddot{\boldsymbol{Z}}(k+1) - \ddot{\boldsymbol{Z}}(k)\right]. \tag{A4.2}$$

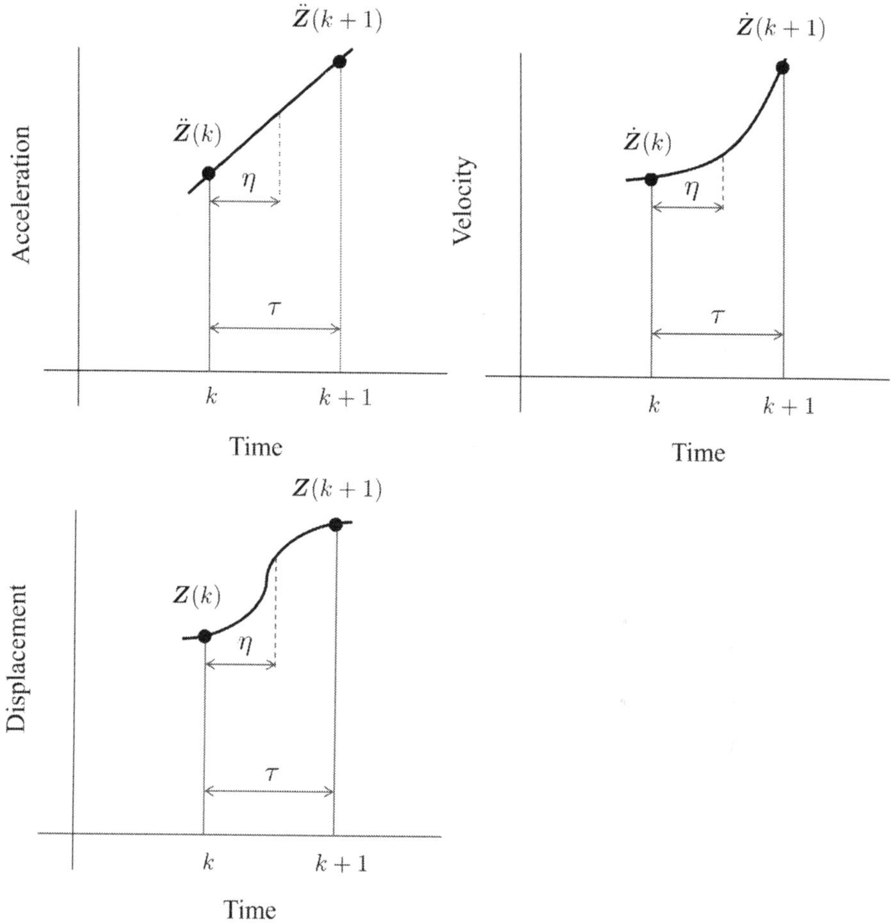

FIGURE A4.1 Newmark β method.

The velocity $\dot{\boldsymbol{Z}}_\tau$ and the displacement $\boldsymbol{Z}_\tau$ are obtained by integrating the above equation with respect to η as follows:

$$\dot{\boldsymbol{Z}}_\eta = \dot{\boldsymbol{Z}}(k) + \ddot{\boldsymbol{Z}}(k)\,\eta + \frac{1}{2\tau}\left[\ddot{\boldsymbol{Z}}(k+1) - \ddot{\boldsymbol{Z}}(k)\right]\eta^2 , \tag{A4.3}$$

$$\boldsymbol{Z}_\eta = \boldsymbol{Z}(k) + \dot{\boldsymbol{Z}}(k)\,\eta + \frac{1}{2}\ddot{\boldsymbol{Z}}(k)\,\eta^2 + \frac{1}{6\tau}\left[\ddot{\boldsymbol{Z}}(k+1) - \ddot{\boldsymbol{Z}}(k)\right]\eta^3 , \tag{A4.4}$$

where $\eta = \tau$. Because $\ddot{\boldsymbol{Z}}_\eta = \ddot{\boldsymbol{Z}}(k+1)$, $\dot{\boldsymbol{Z}}_\eta = \dot{\boldsymbol{Z}}(k+1)$ and $\boldsymbol{Z}_\eta = \boldsymbol{Z}(k+1)$, we obtain

$$\dot{\boldsymbol{Z}}(k+1)=\dot{\boldsymbol{Z}}(k)+\frac{1}{2}\tau\left[\ddot{\boldsymbol{Z}}(k)+\ddot{\boldsymbol{Z}}(k+1)\right], \tag{A4.5}$$

$$\boldsymbol{Z}(k+1)=\boldsymbol{Z}(k)+\tau\dot{\boldsymbol{Z}}(k)+\frac{1}{3}\tau^2\,\ddot{\boldsymbol{Z}}(k)+\frac{1}{6}\tau^2\,\ddot{\boldsymbol{Z}}(k+1). \tag{A4.6}$$

The Newmark β method assumes that the velocity $\dot{\boldsymbol{Z}}(k+1)$ and displacement $\boldsymbol{Z}(k+1)$ at time $k+1$ are given by the following equations, with β representing the change in acceleration within the time interval:

$$\dot{\boldsymbol{Z}}(k+1)=\dot{\boldsymbol{Z}}(k)+\frac{1}{2}\tau\left[\ddot{\boldsymbol{Z}}(k)+\ddot{\boldsymbol{Z}}(k+1)\right], \tag{A4.7}$$

$$\boldsymbol{Z}(k+1)=\boldsymbol{Z}(k)+\tau\dot{\boldsymbol{Z}}(k)+\left(\frac{1}{2}-\beta\right)\tau^2\,\ddot{\boldsymbol{Z}}(k)+\beta\tau^2\,\ddot{\boldsymbol{Z}}(k+1). \tag{A4.8}$$

Substituting Equations (A4.5) and (A4.8) into Equation (A4.1) at time $k+1$ and solving for the acceleration vector $\boldsymbol{Z}(k+1)$ yields

$$\begin{aligned}\ddot{\boldsymbol{Z}}(k+1)=&\left[\boldsymbol{M}+\frac{1}{2}\tau\boldsymbol{C}+\beta\tau^2\boldsymbol{K}\right]^{-1}\Big[\boldsymbol{D}\dot{\boldsymbol{r}}(k+1)+\boldsymbol{E}\boldsymbol{r}(k+1)\\&-\boldsymbol{C}\left[\dot{\boldsymbol{Z}}(k)+\frac{1}{2}\tau\ddot{\boldsymbol{Z}}(k)\right]-\boldsymbol{K}\left[\boldsymbol{Z}(k)+\tau\dot{\boldsymbol{Z}}(k)+\left(\frac{1}{2}-\beta\right)\tau^2\,\ddot{\boldsymbol{Z}}(k)\right]\Big].\end{aligned} \tag{A4.9}$$

When Equation (A4.9) is calculated, the external forces, displacements, velocities and accelerations at time k have already been determined, so all the right-side terms are known values. Therefore, the acceleration at time $k+1$ is obtained and the future response can be calculated sequentially.

Discretising Equation (A4.1) with a sampling period τ using the Newmark β method yields the following state and measurement equations:

$$\boldsymbol{x}(k)=\boldsymbol{F}\boldsymbol{x}(k-1)+\boldsymbol{G}\boldsymbol{u}(k-1)+\boldsymbol{w}(k-1), \tag{A4.10}$$

$$\boldsymbol{z}(k)=\boldsymbol{H}(k)+\boldsymbol{v}(k). \tag{A4.11}$$

where $\boldsymbol{x}(k)$ is the state vector, $\boldsymbol{u}(k)$ is the input vector, $\boldsymbol{z}(k)$ is the measurement, $\boldsymbol{w}(k-1)$ is the system noise and $\boldsymbol{v}(k)$ is the sensor noise. Also, $\boldsymbol{F}$ is the state transition matrix, $\boldsymbol{G}$ is the input matrix and $\boldsymbol{H}$ is the measurement matrix.

The state transition matrix $\boldsymbol{F}$ and the input matrix $\boldsymbol{G}$ can be obtained as

$$\boldsymbol{F} = \boldsymbol{A}^{-1}\boldsymbol{B}, \tag{A4.12}$$

$$\boldsymbol{G} = \boldsymbol{A}^{-1}\boldsymbol{C}, \tag{A4.13}$$

where

$$A = \begin{bmatrix} \boldsymbol{I} & \boldsymbol{0} & -\beta\tau^2\boldsymbol{I} \\ \boldsymbol{0} & \boldsymbol{I} & -\frac{\tau}{2}\boldsymbol{I} \\ \boldsymbol{K} & \boldsymbol{C} & \boldsymbol{M} \end{bmatrix},$$

$$\boldsymbol{B} = \begin{bmatrix} \boldsymbol{I} & \tau & \left(\frac{1}{2}-\beta\right)\tau^2\boldsymbol{I} \\ \boldsymbol{0} & \boldsymbol{I} & \frac{\tau}{2}\boldsymbol{I} \\ \boldsymbol{0} & \boldsymbol{0} & \boldsymbol{0} \end{bmatrix},$$

$$\boldsymbol{C} = \begin{bmatrix} \boldsymbol{0} & \boldsymbol{0} \\ \boldsymbol{0} & \boldsymbol{0} \\ \boldsymbol{E} & \boldsymbol{D} \end{bmatrix}.$$

where $\boldsymbol{I}$ is the identity matrix.

NOTES

1 Use $\partial \mathrm{Tr}(\boldsymbol{ABA}^{\mathrm{T}})/\partial \boldsymbol{A} = 2\boldsymbol{AB}$, where $\boldsymbol{B}$ is a symmetric matrix.

2 Use the matrix inversion lemma, $(\boldsymbol{A} + \boldsymbol{BC})^{-1} = \boldsymbol{A}^{-1} - \boldsymbol{A}^{-1}\boldsymbol{B}(\boldsymbol{I} + \boldsymbol{CA}^{-1}\boldsymbol{B})^{-1}\boldsymbol{CA}^{-1}$.

3 It also can be denoted as $\boldsymbol{P}(k|k) = \boldsymbol{P}(k|k-1) - \boldsymbol{K}(k)\boldsymbol{S}(k|k-1)\boldsymbol{K}(k)^T$.

REFERENCES

1. D. Simon, *Optimal State Estimation: Kalman, H Infinity, and Nonlinear Approaches*, Wiley-Interscience, 2006.
2. F.L. Lewis, L. Xie, D. Popa, *Optimal and Robust Estimation, With an Introduction to Stochastic Control Theory*, CRC Press, 2007.

Index

P

R

S

T

V

W

Y

Made in the USA
Las Vegas, NV
19 September 2026

4a8e7432-963f-413e-8de5-ac661ec24ea0R01